An Introduction to the Electron Theory of Solids

An Introduction to the
Electron Theory of Solids

BY

JOHN STRINGER, B.ENG., PH.D., A.I.M.
Professor of Physical Metallurgy, University of Liverpool

PERGAMON PRESS
OXFORD · LONDON · EDINBURGH · NEW YORK
TORONTO · SYDNEY · PARIS · BRAUNSCHWEIG

Pergamon Press Ltd., Headington Hill Hall, Oxford
4 & 5 Fitzroy Square, London W.1

Pergamon Press (Scotland) Ltd., 2 & 3 Teviot Place, Edinburgh 1

Pergamon Press Inc., 44–01 21st Street, Long Island City, New York 11101

Pergamon of Canada, Ltd., 6 Adelaide Street East, Toronto, Ontario

Pergamon Press (Aust.) Pty. Ltd., 20–22 Margaret Street,
Sydney, New South Wales

Pergamon Press S.A.R.L., 24 rue des Écoles, Paris 5e

Vieweg & Sohn GmbH, Burgplatz 1, Braunschweig

First edition 1967

Library of Congress Catalog Card No. 66–29606

Printed in Great Britain by Bell and Bain Ltd., Glasgow.

(3139/67)

Contents

Introduction

This book is based on a course developed for the express purpose of teaching some solid state theory to students of electrical engineering at the University of Liverpool. The whole course covered three years: in the first year a general introduction to the properties of matter was given. This involved a description in fairly simple terms of the properties of gases, liquids, insulators, semiconductors and metals. The simple Bohr model was described, and from this a qualitative description of chemical binding was developed. Finally, an introduction to crystallography was given, so that the student was familiar with the concepts of macroscopic symmetry and the use of Miller Indices. The idea of a crystal as a three-dimensional diffraction grating was also introduced at this stage, and the Bragg law was derived.

The student during this first year also studied mathematics, and in particular treated vector algebra in some detail. In addition, partial differential equations were introduced, and the idea of real and complex variables. However, his feeling for differential equations had not reached the stage where the appropriate solution is evident on inspection.

This summarises the general background that was assumed in preparing the second-year course on which this book is based. As a consequence, there is no hesitation in introducing differential equations, although so far as possible the solutions are derived in detail. The concept of vectors has been used, and on occasions simple vector algebra: but no vector calculus.

The engineering student adopts a pragmatic attitude to his studies, and the relevance of theoretical manipulations to real problems is a matter of some importance to him. The course as given differed slightly from the book for this reason, since in a textbook logical development takes precedence over the maintenance of student interest, which is a task for the teacher. However, the choice of a historical development, and the unusually great emphasis placed on the experimental grounds which led to the Uncertainty Principle are adopted for this reason. Similarly, in many places physical models are given to illustrate quantum mechanical results, although stress is always laid on the impossibility of accurately modelling quantum situations in a classical world.

In some places, the treatments given in the book are far more detailed than they were in the course. For example, the hydrogen atom is treated in as much detail as the student's mathematical knowledge will permit, although in the course only the first and second quantum numbers were covered in any detail, the remaining two being discussed qualitatively. This is because students often wished to read further in specific areas than the course required. The extended sections correspond to those parts of the course in which more information was most frequently sought.

The final year of the course dealt with the application of the theoretical concepts to real problems in semiconductor devices. This part of the course, given by Dr. J. H. Leck, is also published by Pergamon under the title, *Theory of Semiconductor Junction Devices*.

Because the course was designed for students eventually concerned with semiconductor devices, much more stress has been laid on these aspects of solid state theory; and for example the problem of metallic conduction is treated only cursorily. Problems of interest to the physical metallurgist, such as alloy theory, are also neglected. However, a course very similar to the one covered in this book was given for some years in the Department of Metallurgy at the University of Liverpool as an introduction to a more advanced course on the properties of metals. It is felt therefore that the book is a useful introduction to the theory of electrons in solids for students of any discipline, and introduces those concepts which are needed for more specialised studies of specific classes of materials.

A final obvious omission in a book of this sort is any mention of magnetism. This does not imply that a student of electrical engineering, or any student of the properties of solids, would not be interested in magnetism, or that useful statements cannot be made at this relatively elementary level. However, the immediate aim of the book is to introduce concepts which will aid in an understanding of semiconductors and simple transport processes in solids. For a study of magnetism to assist in this aim, it would have to be more complex (and occupy more space and time) than its importance at this stage warrants. It was therefore decided to omit any discussion of magnetism.

Any lecture course, and any textbook, relies greatly on earlier books. It is impossible to list all the sources one refers to in preparation without becoming cumbersome: but I owe more obvious debts to *Atomic Theory* by Born; *Fundamental Principles of Quantum Theory* by Heisenberg; *Atomic Theory for Students of Metallurgy* by Hume-Rothery; *Electrons and Holes in Semiconductors* by Shockley; *Physical Chemistry* by Moelwyn-Hughes; *Theory of Metals* by Wilson; *Properties of Metals and Alloys* by Mott and Jones; *Solid State Physics* by Kittel; *Elements of Wave Mechanics* by Mott; *Quantum Mechanics* by Rojansky; *Electrons and Phonons* by Ziman; *Theory of the Chemical Bond* by Pauling; *The Electrical and Magnetic Properties of Solids* by Cusack; *Semiconductor Statistics* by Blakemore; *Waves in Periodic Solids* by Brillouin; and *Atomic Spectra* by White. I feel that a student would profit from reading the books by Born, Heisenberg and Hume-Rothery while taking this course; and the book by Kittel, and particularly that by Cusack, are excellent for students who wish to continue their studies further.

CHAPTER 1

The Breakdown of Classical Theory

Particles versus Waves

At the end of the seventeenth century Newton developed the basic ideas of classical mechanics. These well-known laws depend on the assumption that a dynamical system possesses at any instant clearly defined values of the dynamical variables, such as momentum, position, configuration, and so forth; and that there exists—at least in principle—an ideal experiment or set of experiments which can measure these quantities with unlimited accuracy. Because of these assumptions, the need for an experimental determination of the dynamical variables does not explicitly appear in Newtonian mechanics. This essentially Platonic view of mechanics is justified by experience in the case of macroscopic systems, that is, those that can be observed with the naked eye.

Newton suggested that light consisted of a stream of small particles, which he named corpuscles, subject to the same mechanical laws as macroscopic systems. This model is capable of explaining the laws of geometric optics.

In 1690 Huyghens suggested that light was in fact an undulatory displacement of some medium, analogous to the undulatory compression and rarefaction associated with the propagation of sound. However, since light is propagated through a vacuum it was necessary to postulate the existence of an imponderable aether permeating all space. This theory also is capable of explaining the laws of geometric optics, but since the postulation of an aether is not required for the corpuscular theory, the wave theory was largely discounted during the eighteenth century.

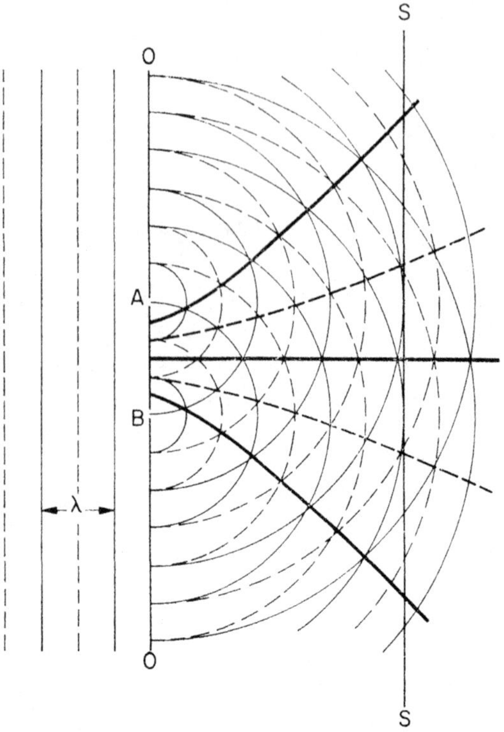

FIG. 1.1. Young's experiment. A plane wave of light, wavelength λ, falls on an opaque screen O, containing two parallel slits at A and B. Cylindrical wavefronts originate at the two slits. In the figure, the maxima at a time t are shown as full lines, the minima as dashed lines. Positions where two maxima or minima coincide clearly result in a combined intensity equal to the sum of that from the two slits. Conversely, a position where a maximum coincides with a minimum corresponds to an intensity which is the difference between that from the slits. The heavy full and dashed lines are thus the locus of maximum and minimum intensity, respectively.

However, early in the nineteenth century Young demonstrated that in certain circumstances two beams of light may "interfere" with each other, the total intensity of the two beams being less than the sum of the separate intensities. This is wholly incompatible with the corpuscular theory, but is easily explained by the wave theory provided that one assumes that the light waves may be superposed so that the aether displacement at any point in space due to the combined waves is the algebraic sum of the displacements due to the two waves separately. Thus, if two waves of the same wavelength and amplitude but 180° out of phase are superimposed, the resultant wave has zero amplitude everywhere, producing complete extinction.

Young's experiment is shown in Fig. 1.1. A parallel monochromatic beam of light falls on an opaque screen O in which are two parallel fine slits A and B. From each of these slits a cylindrical wavefront spreads outwards. The amplitude at any point in the image space to the right of the screen O is then the algebraic sum of the amplitudes of the two component waves at that point. Since the experimental arrangement determines that the amplitudes are of the same magnitude and are in phase at the two slits, the amplitude (and hence the intensity of the light) can be calculated at any point from simple geometry. Suppose a screen is placed at a position S. Then the intensity at a point X on the screen will be zero if the path difference $(AX-BX)$ is equal to an odd number of half-wavelengths, and equal to the sum of the two separate intensities if the path difference is an even number of half-wavelengths. The image on the screen is thus composed of alternating dark and light bands parallel to the slits. If the distance AB is fairly small compared to the distance OS (a condition not satisfied in Fig. 1.1), then near the centre of the screen the bands will be approximately evenly spaced, the distance between the centres of two light bands being $\lambda OS/AB$, where λ is the wavelength of the light.

If one of the slits is covered, the diffraction pattern alters completely. The image on the screen now consists of a central line, directly opposite to the open slit, with much fainter satellite lines on either side. This pattern can be interpreted as arising from the interference of the cylindrical wavefronts generated at the edges of the slit. The diffraction pattern of the pair of slits is thus not a superimposition of two patterns arising from the two separate slits, but necessarily involves the complete geometry: the simultaneous participation of both slits is required to produce the observed diffraction pattern. Since the classical corpuscle must plainly pass through one slit or the other, and not both, this result also is incompatible with the Newtonian model. During the nineteenth century, consequently, the corpuscular theory of light was wholly discounted, and the wave theory of light was developed. In particular Maxwell demonstrated that light was one part of a complete spectrum of electromagnetic waves, and so completely did the predictions of the wave theory agree with experiment that in 1899 the physicist H. Hertz remarked "the wave theory of light is, humanly speaking, a certainty".

The Photoelectric Effect

This remark was made just before the classical wave theory of light was destroyed, and, curiously enough, Hertz had discovered in 1887 the effect which was to disprove it. This is the photoelectric effect. When light falls on a clean metal surface in vacuo, electrons are emitted from the surface, and may be collected and appear as a current in an external circuit. The current is then a measure of the number of electrons reaching the collector. If a grid is interposed between the emitter and the collector, the energy distribution of the emitted electrons can be measured by observing the variation of the current with the grid bias. The circuit is shown in Fig. 1.2. In particular, the maximum energy of the emitted electrons is measured by the field required to reduce the current to zero. It is found that (i) the number of electrons emitted in unit time is proportional to the intensity of the illumination, (ii) the maximum energy of the electrons is *not* dependent on the intensity of the light, but only on the frequency, and (iii) there exists a cut-off frequency ν_0, so that light for which $\nu < \nu_0$ does not excite emission at all, no matter how great the intensity.

These observations are incompatible with the wave theory of light, since this theory postulates that the energy flux at the surface of the metal is uniform. Since emission of an energetic electron is caused by the irradiation, presumably the metal must store the energy in some way, finally discharging it all to an electron. However, in this case the energy of the electron should be determined by the storage process, which is presumably

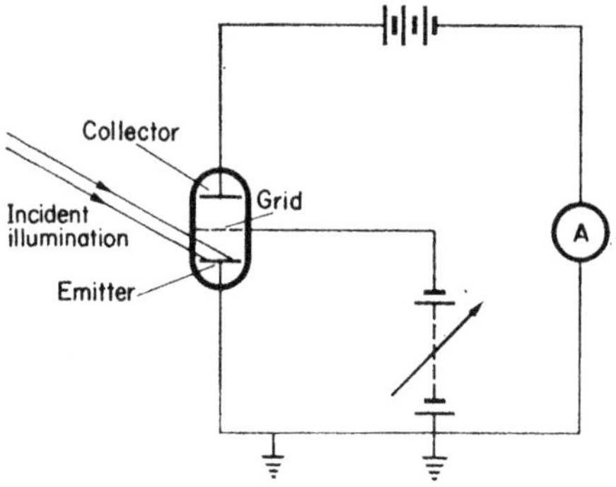

FIG. 1.2. The photoelectric effect.

a property of the metal, and possibly also by the energy flux. Plainly also there should be no cut-off frequency in this model.

For light of high frequency, the emitted electrons have high energy. If the intensity of the illumination is reduced sufficiently, the emission of an electron from the surface is an infrequent occurrence. This suggests that the storage mechanism does not "leak" at all, since the energy of the electron in this case is the same as that when the illumination is more intense and the storage time shorter. Consequently, one would expect metal to exhibit a memory of previous irradiation, so that emission should commence earlier on a previously irradiated piece of metal. In fact this is not so. An elegant experiment to demonstrate this point was performed by Meyer and Gerlach in 1914.

The emitter was a finely divided metal powder, so that the small grains would each require a long time to accumulate sufficient energy to produce emission from a continuous wave. For the experimental conditions used by Meyer and Gerlach there should thus have been a delay of some seconds before emission commenced, but in fact emission always commenced immediately the powder was illuminated. It therefore appears that emission does not involve energy storage by the metal; but in this case there is no way of explaining the results on the basis of the wave theory of light.

Einstein suggested in 1905 that in fact the energy in the light wave does not arrive continuously at the metal surface, but discontinuously in "packets" which are indivisible. These packets, or quanta, we shall call "photons". The phenomenon of emission is then a simple collision process, a photon colliding with an electron in the metal and giving it sufficient energy to escape. There is thus no storage process. Einstein further postulated that the energy of the photon is related to the frequency of the light as measured in an interference experiment by the equation

$$E = h\nu \tag{1.1}$$

where h is a constant. This equation is called the Einstein relationship.

Let the quantity of energy required to remove an electron from inside the metal to a position at rest immediately outside be ϕ. Then the absorption of a photon energy E causes the emission of an electron with a kinetic energy T where

$$E = T+\phi. \tag{1.2}$$

Since the electron may lose energy after the absorption of the photon by collisions with atoms in the surface of the solid, T will be the maximum energy in the emitted electrons. If the grid bias required to reduce the collector current to zero is $\mathscr{E}$, then

$$T = e\mathscr{E} \tag{1.3}$$

where e is the charge on the electron. Substituting (1.1) and (1.3) in (1.2) we obtain

$$h\nu = e\mathscr{E} - \phi. \tag{1.4}$$

It follows that if the frequency ν of the illumination is varied and the stopping field $\mathscr{E}$ measured for a number of metal surfaces, (1.4) predicts that a graph of ν versus $\mathscr{E}$ should be a straight line with a slope that is a universal constant, but with an intercept on the $\mathscr{E} = 0$ axis of

$$h\nu_0 = \phi. \tag{1.5}$$

Plainly this will be a property of the metal surface. ϕ is called the photoelectric work function. For frequencies $\nu < \nu_0$, no emission will take place, so that ν_0 is a cut-off frequency.

The predictions of (1.4) have been carefully checked by Richardson and Compton (1912), by Hughes (1912) and by Millikan (1916) amongst others, and confirmed in detail.

Einstein's suggestion of the quantisation of radiation was an extension of earlier work by Planck (1900) on the black-body radiation of heat, another problem which could not be treated successfully using classical laws. Planck suggested that heat is radiated by material oscillators in the black body, and that the energy of these oscillators could have only one of the values

$$E = nh\nu \tag{1.6}$$

where n is an integer. The radiation of heat by these oscillators then corresponds to a jump from one of these energy levels to the next lower. With this hypothesis, Planck was able to calculate accurately the distribution of frequencies in the emission spectrum from a black body. The value of the constant h obtained from the radiation of heat corresponds very closely to that obtained from the photoelectric effect. h is called Planck's constant, and it is plain from (1.1) that it has dimensions of action. The presently accepted value is

$$h = 6{\cdot}6253 \times 10^{-27} \text{ erg sec.} \tag{1.7}$$

Although Einstein's hypothesis (1.1) appears to be a simple extension of Planck's hypothesis (1.6), in fact, as we shall see, it embodies an important new idea.

How Big is a Photon?

In 1925 Joffé and Dobronravoff attempted to measure the lateral area of a photon. For reasons of experimental convenience, they chose to study a photon from the X-ray region of the electromagnetic spectrum rather than the optical region. The anticathode of their X-ray tube was an aluminium plate which also formed one plate of a capacitor. Into the electrostatic field of this capacitor was introduced a small particle of bismuth, which on absorption of an X-ray photon emitted an electron and thus became positively charged. In a manner similar to the Millikan oil-drop experiment, the charged particle could be supported in space by balancing the effects of gravity and the electrostatic field. Each time the particle absorbed a photon, its charge altered, and consequently its motion in the field altered: these changes were observed with a telescope. At the X-ray source, the particle, which was between 1 and 5×10^{-5} cm in diameter, subtended a solid angle of less than 10^{-5}. Within this angle, the tube emitted one photon every 30 min. In fact the charge on the particle altered only once every 30 min, and from this it can be concluded that the particle absorbed only photons emitted within its own solid angle; that is, photons whose centres would have struck it. The radial extent of a photon normal to the direction of propagation is thus small in comparison to the diameter of the particle of bismuth.

Lawrence and Beams in 1928 attempted to measure the longitudinal extent of a photon, constructing a shutter capable of opening for periods as short as 10^{-9} sec. This consisted of two crossed Nicol prisms with two Kerr cells at right angles between them. A Kerr cell consists of two capacitor plates with a suitable organic liquid dielectric between them, such as carbon disulphide, chloroform or nitrobenzene. An electric field across the capacitor causes light passing through the dielectric to be elliptically polarised. So long as the field in the two Kerr cells is the same, one compensates the other, and the crossed Nicols ensure that no light is passed. If, however, the field in the two cells is different, light is passed through the assembly. The two cells are charged in parallel, with a different length of wire attached to each. Consequently, for a short time determined by the speed of the current along the additional length of wire the charges in the two cells are not equal, and light is passed.

This very rapid shutter did permit photons to pass through, and failed to cut any photons, confirming that photons appear to be indivisible. The longitudinal extent of a photon is thus less than a few centimetres.

Wave–Particle Duality

The photoelectric effect and the later experiments suggest that light is not a continuous wave, as suggested by the classical wave theory, but instead is a stream of photons, indivisible, with a well-defined energy, and small in extent. These are close to the properties of a classical particle, so that it now seems appropriate to question again the interference experiment of Young, to inquire whether possible interactions between photons in the image space may produce the interference pattern. G. I. Taylor performed a crucial experiment in 1909, repeating the twin-slit interference experiment with light of such low intensity that there was only one photon in the optical system at any time. The detector used was a photographic plate, which was left exposed for a long time, so that the total number of photons arriving at the plate was similar to the number arriving in a short time at normal intensities. The resultant pattern on the plate was exactly the same as the pattern seen on the screen at normal intensities, as in Young's experiment. However, it has already been demonstrated that the interference pattern requires the simultaneous presence of both slits. Consequently, we are forced to the conclusion that each photon (which we have demonstrated experimentally is very small and indivisible) has nevertheless passed through both slits at once. This paradox demonstrates that in fact light is *neither* a stream of classical particles *nor* a classical wave. We therefore accept that light will exhibit in some experiments particle-like properties, and in other experiments wave-like properties, and we call this "duality", although it must be noted that the duality exists only in the inappropriateness of our descriptions, and not in the light itself. It is important to note that the experiments of G. I. Taylor prove that duality is a property that extends down to the finest scale, that of single photons, and is not in any sense a statistical property of large assemblies of photons.

Although in this review of the development of the concept of duality we have concentrated on the visible part of the electromagnetic spectrum the general conclusions are valid for the whole spectrum. It is easy to demonstrate the dual nature of the short wave end—the existence of X-ray photons are demonstrated for example by cloud chamber experiments or by experiments like that of Joffé and Dobronravoff. The wave nature is demonstrated by diffraction experiments, using a crystal as a diffraction grating, as in the experiments of Bragg, or a ruled grating, as in the experiments of Compton. For the long wavelength end of the spectrum the particle nature is more difficult to demonstrate, because of the very low energy of individual photons. Thus, for light $\lambda \simeq 6000$ Å, so $\nu \simeq 0{\cdot}5 \times 10^{15}$ sec^{-1}, so that from (1.1) the energy of a photon of visible light is approximately $3{\cdot}3 \times 10^{-12}$ erg, or 2 eV, where 1 eV (electron volt) is the energy acquired by an electron falling through a potential difference of 1 V. This energy is comparable with that required to excite various reactions, so that it can be easily observed experimentally. On the other hand a radio wave with a wavelength of 1 cm and consequently a frequency of 3×10^{10} sec^{-1}, has a photon energy of approximately 10^{-4} eV. This is close to the energy acquired by a classical kinetic theory gas molecule heated by one degree absolute, and consequently the detection of a single photon presents considerable experimental problems. In general, we note that as the wavelength becomes shorter the particle nature becomes more pronounced, and the wave character less so.

This new and startling concept of duality is embodied in the Einstein relation (1.1), which is why it differs in an important respect from the earlier postulate of Planck. However, physicists abandoned the old ways of thought with reluctance, general acceptance of the existence of duality taking some twenty years after Einstein's original statement.

The Particle Character of the Electron

The electron, unlike a photon, was thought of as a particle from soon after its discovery. Also unlike a photon it possesses a rest mass, and is plainly not part of the electromagnetic spectrum. There are a large number of experiments which demonstrate the particle nature of the electron, indicating that it has a well-defined rest mass, a well-defined charge and a limited physical extent. A typical experiment is that of Millikan (1910), in which fine oil drops were injected into the air space between two capacitor plates, and their terminal velocity of fall under gravity determined with a telescope. From the terminal velocity their size was calculated, using Stokes' Law. The gas in the apparatus was then irradiated with X-rays, ionising the gas molecules, producing electrons and positively charged ions. These were adsorbed on the oil droplets. Application of an electric field between the capacitor plates modified the rate of fall of the charged droplets. By measuring the new terminal velocity the charge on the droplets was determined. This always turned out to be an integral multiple of a basic unit, the charge on a single electron. The value of the charge determined from this experiment and similar experiments is

$$e = -4{\cdot}803 \times 10^{-10} \text{ e.s.u.} \tag{1.8}$$

The electron may be deflected by magnetic fields, and the deflection produced permits a calculation of the charge-to-mass ratio of the electron. Using the above value of the charge, the mass of an electron may be calculated. The currently accepted value is

$$m = 9{\cdot}108 \times 10^{-28}\ \text{g}. \tag{1.9}$$

Collision experiments suggest that the effective diameter of the electron is about 10^{-13} cm. The electron is thus very close to the Newtonian concept of a particle, and in fact Newtonian mechanics describes the motion of a particle in free space very accurately. However, in 1925 Elsasser suggested that electrons might be diffracted by a crystal lattice in the same way as photons, and in 1927 Davisson and Germer demonstrated that this was in fact so. Nowadays, electron diffraction and X-ray diffraction are both powerful and commonly used experimental tools for the elucidation of crystal structure. It follows consequently that duality is not a unique property of the electromagnetic spectrum, but apparently applies to material particles (those having a non-zero rest mass) as well. This general result was predicted on theoretical grounds by Louis de Broglie in 1924, who predicted that the wavelength associated with the material particle would be related to its momentum p by the equation

$$\lambda = \frac{h}{p}. \tag{1.10}$$

This prediction is borne out exactly by experiment.

The existence of a wave character in material particles, and the consequent applicability of the concept of duality to them, has since been generalised to more massive particles: neutrons, alpha particles, even atoms of lighter elements have exhibited wave character in experiments, and the quantitative results agree exactly with the Einstein relationship (1.1) and the de Broglie relationship (1.10).

CHAPTER 2

Atomic Spectra and the Old Quantum Theory

The Spectrum of Atomic Hydrogen

The hydrogen atom consists of a positively charged nucleus, and an electron with an equal negative charge. The diameter of the atom as a whole is about 10^{-8} cm, and collision experiments show that both the nucleus, which is called a proton, and the electron have diameters of about 10^{-13} cm. This means that the hydrogen atom is very much more empty than the solar system: the radius of the atom compared to that of the nucleus is about 10 times that of the orbit of Pluto compared to the sun. If the atom were magnified to a diameter of one metre, the radius of the electron and the nucleus would each be only about one hundredth of a millimetre. It follows that from the classical point of view both components may be regarded as point masses and charges. The force of attraction between them is the coulomb electrostatic attraction, and this suggests on classical reasoning a planetary atom, with the electron and proton orbiting around their common centre of gravity. Since the proton mass is approximately 1840 times that of the electron, the centre of gravity is close to that of the proton.

If the atom is bombarded with electrons or X-ray photons, its energy increases and it is said to be excited. In terms of the planetary model, the electron increases its orbital radius. If sufficient energy is supplied, the separation between the proton and the electron may approach infinity. The atom is then said to be ionised, and the energy required to remove the electron from the lowest energy state of the atom to a position at rest at infinity is called the first ionisation potential. We choose this condition as the zero of energy, so that when the electron is "bound" to the atom, i.e. when its energy is insufficient for it to escape from the proton, the energy is negative.

The planetary problem has been analysed by Kepler, who demonstrated that the possible orbits are conic sections, bound orbits corresponding to circles and ellipses, and unbound orbits (positive energy) corresponding to parabolae and hyperbolae. The former orbits are periodic, the orbiting body returning to a specific position at regular intervals of time: the unbound orbits are, however, not periodic, each orbit being a unique event. From the point of view of the classical planetary theory no orbit is inherently more stable than any other. If a mechanism exists for radiating energy, the energy of the system will diminish until the orbiting body falls into its primary. In addition, any increment of energy may be accepted by the system. The experimental results for the hydrogen atom are quite at variance with this for the bound orbits, although for unbound orbits it is true that the energy may be altered continuously. In the negative energy region, the atoms can only accept or emit specific amounts of energy. Excited atoms, for example, emit energy discontinuously in the form of photons with specific well-defined frequencies, until the atom reaches a particular value of the energy, called the ground state, in which it appears to have infinite stability, since no further emission takes place. The radiation emitted by excited atomic hydrogen does not therefore form a continuous spectrum, as the classical model predicts, but a "line spectrum" corresponding to the negative energy region, together with a "continuous spectrum" adjoining it corresponding to the positive energy region.

The emission of a line spectrum is not in itself entirely opposed to classical concepts. The sound waves emitted from a violin string form a line spectrum, but in this case all the terms can be represented as harmonics of a fundamental. The line spectrum of atomic hydrogen cannot be described in this way. In 1885

Balmer demonstrated that the visible part of the spectrum was described by the equation

$$\frac{1}{\lambda} = R_H \left(\frac{1}{4} - \frac{1}{m^2}\right) \tag{2.1}$$

where m is an integer greater than 2. R_H is a constant, called the Rydberg constant for atomic hydrogen. It soon became clear that (2.1) was one form of a more general equation:

$$\frac{1}{\lambda} = R_H \left(\frac{1}{n^2} - \frac{1}{m^2}\right) \tag{2.2}$$

where n, m are both integers, with $m > n$. For each value of n we generate a series of lines, of which (2.1) is called the Balmer series. The names of the principal series in the spectrum of atomic hydrogen are given in Table 2.1.

TABLE 2.1

n	Series	Location	Date
1	Lyman	Ultraviolet	1915
2	Balmer	Visible	1885
3	Paschen	Near infrared	1908
4	Brackett	Far infrared	1922
5	Pfund	Far infrared	1924
6	Humphreys	Far infrared	1953

Formula (2.2) is a particular example of a principle noted by Ritz in 1908. This principle, called the Ritz–Rydberg combination principle, states that all the lines in a given spectrum are related by a "term series" such that the reciprocal of the wavelength of each line is the difference between two "terms". Examination of (2.2) shows that the term series for atomic hydrogen is

$$R_H; \quad R_H/4; \quad R_H/9; \quad \ldots; \quad R_H/n^2; \quad \ldots; \quad 0. \tag{2.3}$$

The Old Quantum Theory

In 1913 Bohr suggested that instead of the energy of the electron orbits being continuously variable, as in the classical model, there existed certain orbits with high stability, in the sense that the electron could remain in these orbits for finite times without radiating. These orbits we call "stationary states". Every spontaneous change from a stationary state of the system results in a transition to another stationary state, and the emission of a photon. If the energy of the initial state is E_1 and that of the final state E_2, the frequency ν of the emitted photon may be calculated from the Einstein relation (1.1):

$$\Delta E = E_1 - E_2 = h\nu_{12}. \tag{2.4}$$

This process is reversible: the atom can absorb a photon provided the photon energy is sufficient to raise the atom from one stationary state to another.

It follows from this hypothesis of Bohr that the inferred term series (2.3) represents the permitted energies corresponding to the stationary states. These are called "energy levels". The permitted energy levels are thus

$$E_n = h\nu_n = -\frac{hc}{n^2} R_H \tag{2.5}$$

where as usual the negative sign infers that the states are bound. It follows from this that the quantity hcR_H is the first ionisation potential for atomic hydrogen, which is called a Rydberg. The currently accepted value is

$$1 \text{ Rydberg} = 13{\cdot}60353 \text{ eV}. \tag{2.6}$$

The energies of the stationary states as given by (2.5) are shown on the left-hand side of Fig. 2.1. The integer n is called the Principal Quantum Number.

So far, the Bohr theory is simply a restatement of the experimental results in terms of the Ritz–Rydberg principle and the Einstein relation. Bohr now postulates that the electron in the stationary orbit may be treated by classical mechanics, so that the stationary orbits are in fact Kepler orbits. It follows that only the closed periodic orbits correspond to stationary states; the open aperiodic orbits have continuously variable energy. It appears therefore that quantisation is associated with periodicity.

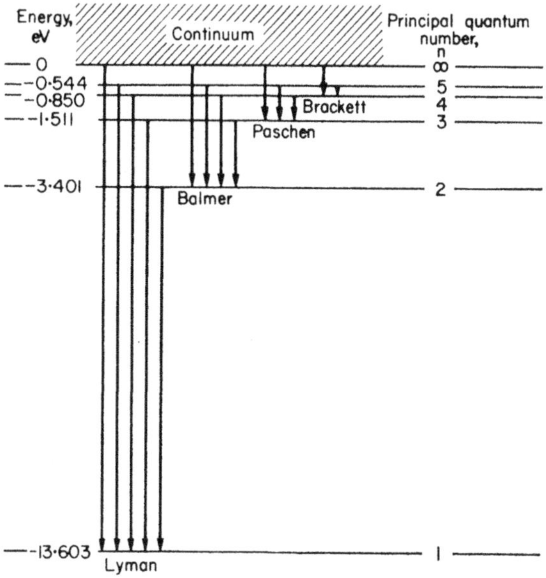

FIG. 2.1. Level diagram for atomic hydrogen.

Suppose that the motion of a general system having a number of degrees of freedom may be described by means of position coordinates q_k and momentum coordinates p_k so chosen that each p_k is a function only of the corresponding q_k. The quantum condition of Bohr may then be written

$$\oint p_k dq_k = n_k h \tag{2.7}$$

where n_k is an integer, and the symbol $\oint$ indicates that the integral extends over a complete cycle of the variable q_k. [This general formulation of the Bohr quantisation condition differs slightly from that of Bohr, and is due to W. Wilson (1915), Sommerfeld (1916) and Ishiwara (1915).]

Now we consider the motion of a system consisting of an electron mass m and charge $-e$ and a proton mass M and charge $+e$. The potential energy is $-e^2/r$, where r is the distance between the proton and the electron. If the position coordinates of the electron and proton respectively are

$$(x_1, y_1, z_1) \quad \text{and} \quad (x_2, y_2, z_2),$$

the energy E is given by

$$E = \tfrac{1}{2}m(\dot{x}_1^2+\dot{y}_1^2+\dot{z}_1^2)+\tfrac{1}{2}M(\dot{x}_2^2+\dot{y}_2^2+\dot{z}_2^2)-\frac{e^2}{r} \tag{2.8}$$

where as usual the superposed dot indicates differentiation with respect to time. If the coordinate system is chosen so that the centre of gravity of the system is at rest, then

$$mx_1+Mx_2 = 0$$

and similarly for y and z. A new set of coordinates (x, y, z) may therefore be introduced, where $x = x_2-x_1$, etc. Then

$$x_1 = \frac{-M}{M+m}x; \quad x_2 = \frac{m}{M+m}x. \tag{2.9}$$

and similarly for y_1, y_2; and z_1, z_2.
The corresponding terms for the kinetic energy are

$$\tfrac{1}{2}m\dot{x}_1^2+\tfrac{1}{2}M\dot{x}_2^2 = \tfrac{1}{2}\mu\dot{x}^2 \tag{2.10}$$

where

$$\mu = \frac{mM}{m+M}$$

is called the reduced mass of the system.

It is convenient here to invoke Kepler's second law, which states that the line joining the two bodies sweeps out equal areas in equal times, and implies that the motion is coplanar. We define this plane as the xy-plane, and transform to polar co-ordinates:

$$x = r\cos\theta; \qquad y = r\sin\theta.$$

The second law of Kepler is equivalent to a statement of the conservation of angular momentum:

$$\mu r^2\dot{\theta} = p_\theta = \text{constant} \tag{2.11}$$

where p_θ is the angular momentum.

The energy becomes

$$E = \tfrac{1}{2}\mu(\dot{r}^2+r^2\dot{\theta}^2)-\frac{e^2}{r} = \text{constant}, \tag{2.12}$$

from the principle of conservation of energy. Eliminating $\dot{\theta}$ between these two equations, and noting that

$$\dot{r} = \dot{\theta}\frac{dr}{d\theta} = \frac{p_\theta}{\mu r^2}\frac{dr}{d\theta}$$

we have

$$E = \frac{1}{2}\frac{p_\theta^2}{\mu}\left\{\frac{1}{r^4}\left(\frac{dr}{d\theta}\right)^2+\frac{1}{r^2}\right\}-\frac{e^2}{r}. \tag{2.13}$$

Rewriting (2.13) in terms of a new variable $\rho = r^{-1}$, we obtain

$$E = \frac{1}{2}\frac{p_\theta^2}{\mu}\left\{\left(\frac{d\rho}{d\theta}\right)^2+\rho^2\right\}-e^2\rho. \tag{2.14}$$

Differentiating (2.14) with respect to θ and dividing by $d\rho/d\theta$ we obtain

$$\frac{d^2\rho}{d\theta^2}+\rho-\frac{e^2\mu}{p_\theta^2} = 0. \tag{2.15}$$

The solution of this second order differential equation is well known and may be written

$$\rho = \frac{e^2\mu}{p_\theta^2} + C\cos(\theta - \theta_0) \tag{2.16}$$

where C and θ_0 are constants. Since the axis from which θ is measured has not been defined, it may be chosen so that $\theta_0 = 0$. For compactness, we write $q = p_\theta^2/e^2\mu$ and $\varepsilon = Cq$; then (2.16) may be written

$$r = \frac{1}{\rho} = \frac{q}{1+\varepsilon\cos\theta}. \tag{2.17}$$

This is the general equation of a conic section. For the orbit to be periodic, r is always finite, so that the denominator must never vanish. This condition is equivalent to restricting $|\varepsilon|$ to values less than unity.

Perihelion, the minimum value of r, is at $\theta = 0$, so

$$r_{\text{min}} = \frac{q}{1+\varepsilon}. \tag{2.18a}$$

Aphelion, the maximum value of r, is at $\theta = \pi$, so

$$r_{\text{max}} = \frac{q}{1-\varepsilon}. \tag{2.18b}$$

The major semi-axis a is given by

$$a = \frac{1}{x}(r_{\text{max}} + r_{\text{min}}) = \frac{q}{1-\varepsilon^2} \tag{2.19a}$$

ε is the numerical eccentricity of the orbit, and so the minor semi-axis b is given by

$$b = \frac{q}{(1-\varepsilon^2)^{\frac{1}{2}}}. \tag{2.19b}$$

Now, substituting (2.16) into (2.14),

$$E = \frac{1}{2}\frac{p_\theta^2}{\mu}\left\{C^2 - \left(\frac{e^2\mu}{p_\theta^2}\right)^2\right\} - \frac{e^4\mu}{p_\theta^2} = \frac{e^2}{2a}. \tag{2.20}$$

Notice that the eccentricity does not appear in the expression for the energy; all orbits with the same major semi-axis have the same energy. Now although there is only one period, it is possible to specify two quantum conditions:

$$\left.\begin{aligned} \oint p_r dr &= n'h, \\ \oint p_\theta d\theta &= kh, \quad n', k \text{ integers.} \end{aligned}\right\} \tag{2.21}$$

Since p_θ is constant, the second condition may be written

$$p_\theta = \frac{h}{2\pi}\,.\,k. \tag{2.22}$$

The first integral is a little more difficult to evaluate.

$$n'h = \oint p_r dr = \oint \mu \dot{r} dr = \oint \mu \frac{dr}{d\theta}\dot{\theta}\frac{dr}{d\theta}d\theta$$

$$= p_\theta \oint \frac{1}{r^2}\left(\frac{dr}{d\theta}\right)^2 d\theta. \tag{2.23}$$

Thus we have to evaluate the integral

$$I = \int_0^{2\pi} \frac{1}{r^2}\left(\frac{dr}{d\theta}\right)^2 d\theta = \int_0^{2\pi} \frac{1}{\rho^2}\left(\frac{d\rho}{d\theta}\right)^2 d\theta$$

$$= \int_0^{2\pi} \frac{\varepsilon^2 \sin^2\theta}{(1+\varepsilon\cos\theta)^2} d\theta \tag{2.24}$$

for a complete revolution. Integrating by parts,

$$I = \left[\frac{\varepsilon \sin\theta}{1+\varepsilon\cos\theta}\right]_0^{2\pi} - \int_0^{2\pi} \frac{\varepsilon\cos\theta}{1+\varepsilon\cos\theta} d\theta = -\int_0^{2\pi} \frac{\varepsilon\cos\theta}{1+\varepsilon\cos\theta} d\theta \tag{2.25}$$

Multiplying (2.24) by -1 and (2.25) by $+2$ and adding, we obtain

$$I = -\int_0^{2\pi} \frac{\varepsilon^2 \sin^2\theta + 2\varepsilon\cos\theta(1+\varepsilon\cos\theta)}{(1+\varepsilon\cos\theta)^2}\,d\theta$$

$$= -\int_0^{2\pi} \frac{1+2\varepsilon\cos\theta+\varepsilon^2\cos^2\theta}{(1+\varepsilon\cos\theta)^2}\,d\theta + \int_0^{2\pi} \frac{1-\varepsilon^2}{(1+\varepsilon\cos\theta)^2}\,d\theta. \tag{2.26}$$

The first of these integrals may be evaluated immediately, and is -2π, the second, using (2.15), may be written

$$\frac{1-\varepsilon^2}{q^2}\int_0^{2\pi} r^2\,d\theta. \tag{2.27}$$

This integral is equal to twice the area of the elliptical orbit, which is $\pi a.b$. Substituting for a and b and collecting terms, the quantum condition becomes

$$n'h = -2\pi p_\theta + p_\theta \frac{1-\varepsilon^2}{q^2}\,2\pi \frac{q^2}{(1-\varepsilon^2)^{\frac{3}{2}}}$$

$$= -hk + \frac{2\pi p_\theta}{\sqrt{(1-\varepsilon^2)}} \tag{2.28}$$

or,

$$\frac{4\pi^2 p_\theta^2}{1-\varepsilon^2} = 4\pi^2 e^2 \mu a = h^2(n'+k)^2.$$

Setting $n = n'+k$, we have

$$a = \frac{h^2 n^2}{4\pi^2 e^2 \mu}. \tag{2.29}$$

So that the energy of the nth state is

$$E_n = -\frac{e^2}{2a} = -\frac{2\pi^2 e^4 \mu}{h^2 n^2} = hcR_H \cdot \frac{1}{n^2} \tag{2.30}$$

where

$$R_H = \frac{2\pi^2 e^4 \mu}{h^3 c}.$$

For the hydrogen atom consequently the energy of the ground state is

$$E_1 = -\frac{2\pi^2 e^4 \mu}{h^2} = 13{\cdot}59935 \text{ eV} \tag{2.31}$$

which compares well with the experimental value (2.6). The Rydberg constant for atomic hydrogen becomes

$$R_H = \frac{2\pi^2 e^4 \mu}{h^3 c} = 109{,}680{\cdot}024 \text{ cm}^{-1} \tag{2.32}$$

compared to the experimental value of 109,678·18 cm^{-1}. The quantity

$$a_1 = \frac{h^2}{4\pi^2 \mu e^2} = 0{\cdot}52947 \text{ Å} \tag{2.33}$$

is called the first Bohr radius for hydrogen.

The Rydberg constant is a function of mass, and so will differ for different isotopes of the same element. Consequently, hydrogen (mass 1) should have a slightly different spectrum to deuterium (mass 2). In fact, this is how the existence of deuterium was first demonstrated by Urey, Brickwedde, and Murphy in 1932. The calculated and observed displacements for the first few terms in the Lyman series are listed in Table 2.2.

From (2.19b), the minor semi-axis becomes

$$b = \frac{h^2}{4\pi^2 e^2 \mu} nk. \tag{2.34}$$

So

$$\frac{b}{a} = \frac{k}{n}. \tag{2.35}$$

Hence, if $k = n$, the orbit is circular. If $k = 0$, the orbit would be a straight line passing through the proton, the so-called "pendulum swing" orbit. In terms of the classical model, this seems impossible: so it appears that the quantum number k is restricted to the range of values 1, 2, 3, ..., n. The integer n, however, solely determines the energy of the state, and in accordance with the empirical theory is called the principal quantum number.

TABLE 2.2. ISOTOPIC DISPLACEMENT OF THE LYMAN SERIES ($n = 1$)

m	$\lambda \times 10^{-8}$ cm (calculated)		$\Delta\lambda \times 10^{-8}$ cm	
	H	D	calculated	observed
2	1215·664	1215·334	0·330	0·330
3	1025·718	1025·439	0·279	0·276
4	972·533	972·269	0·264	0·266
5	949·739	949·481	0·258	0·262
6	937·800	937·545	0·255	0·274
7	930·744	930·491	0·253	0·258

(From Moelwyn-Hughes: p. 115.)

The quantum number k, which refers to the quantisation of the angular component of momentum, is called the angular quantum number. All states having the same n have the same energy: we say that the system is degenerate with respect to the quantum number k.

Sommerfeld's Modification

The linear velocity of the electron is

$$v = r\omega$$

where ω is the angular velocity. For the circular first Bohr orbit, for which n is 1,

$$\omega = \frac{8\pi^3 e^4}{h^3} \mu \qquad (2.36)$$

and, using (2.33),

$$v = \frac{2\pi e^2}{h} = \frac{c}{137{\cdot}06} \tag{2.37}$$

where c is the velocity of light. Because of this, the non-relativistic mechanics used is inadequate, since the relativistic mass change will be sufficient to contribute significantly to the energy. For circular orbits the correction is not important, since the linear velocity is constant: thus for the first Bohr orbit one simply replaces μ by μ', where

$$\mu' = \frac{\mu}{\sqrt{[1-(1/137{\cdot}06)^2]}}. \tag{2.38}$$

But for elliptical orbits the second law of Kepler requires the linear velocity to vary round the orbit, being a maximum at perihelion and a minimum at aphelion; and the difference between the maximum and minimum velocity increases with increasing eccentricity. Consequently, the mass varies during the orbit, the effect of which is to produce an orbit which is no longer exactly elliptical: the increase in the mass as the electron approaches the proton swings the electron a little further round the proton, so that its new path is slightly in advance of its last. The new orbital path consequently resembles a rosette, and to a first approximation the orbit may be regarded as a "precessing ellipse", that is, an elliptical motion superimposed on a general rotation of the major semi-axis, in the same direction as the orbital motion, about the centre of mass.

The problem was analysed quantitatively by Sommerfeld in 1916, who demonstrated that the advance of the perihelion per revolution is

$$2\pi(\gamma^{-1}-1)$$

where

$$\gamma^2 = 1-\frac{1}{k^2}\,5{\cdot}305\times 10^{-5}. \tag{2.39}$$

For an orbit with $k = 1$, this is approximately $1{\cdot}7\times 10^{-5}$ radians per revolution. Since the precession is so small, we may still regard the path around the orbit as elliptical. The energy of the orbits taking into account the relativistic mass change is given by the formula

$$1+\frac{E}{\mu_0 c^2} = \left\{1+\frac{\alpha^2}{[n'+(k^2-\alpha^2)^{\frac{1}{2}}]^2}\right\}^{-\frac{1}{2}} \qquad (2.40)$$

where as before n' is the radial quantum number, μ_0 is the rest reduced mass, and α is the "fine structure constant"

$$\alpha = \frac{2\pi e^2}{hc} = 7{\cdot}2973\times 10^{-3}. \qquad (2.41)$$

From this expression the term series corresponding to (2.30) may be derived:

$$-\frac{E_{n,k}}{hc} = \frac{R_H}{n^2}+\frac{R_H\alpha^2}{n^4}\left(\frac{n}{k}-\frac{3}{4}\right)+\text{terms of order } \alpha^4. \qquad (2.42)$$

This means that the degeneracy of the orbits of the same major semi-axis but different eccentricity is removed by the relativistic mass change, with a resultant splitting in the spectral lines. The magnitude of the splitting is indicated by the second term in (2.42), and since this contains the term α^2 it is plain that the splitting will be very small. In fact, a fine substructure—the "fine structure"—is observed in each of the spectral lines of atomic hydrogen, and the agreement between the experimental spacing of the fine structure term series and (2.42) is excellent. However, all the lines implied by the term series (2.42) are not present. In order to obtain agreement between experiment and theory, it is necessary to postulate that transitions between all the states are not permitted. Exact agreement is obtained if the transitions are restricted to those in which the angular quantum number changes by unity:

$$\Delta k = \pm 1. \qquad (2.43)$$

This is called a selection rule.

The Third Degree of Freedom—the Zeeman Effect

So far, only two quantum numbers have been used to describe the motion, and this is a necessary corollary of the restriction introduced earlier to the effect that the possible orbits are planar. The third degree of freedom is related to the angle between the normal to the orbit plane and some external reference axis. However, in empty space the choice of such an axis would be completely arbitrary, and therefore not meaningful. It follows that in empty space the system is completely degenerate with respect to this third degree of freedom. To put it another way, since there is no way of distinguishing the orientation of the plane of the orbit in empty space, it does not matter how it is oriented. Consequently, every possible orientation in this third dimension must have the same energy; if this were not so, a measurement of the energy would give a measure of the orientation, which has just been shown to be impossible.

Since the electron carries a charge $-e$, its motion round its orbit will produce a magnetic moment. Faraday stated that the magnetic moment produced by a current I flowing in a loop area S is IS. For convenience, the discussion will be restricted to circular orbits. In this case, $S = \pi r^2$, and $I = \omega e/2\pi$ where ω is the angular velocity of the electron. The magnetic moment M is then

$$M = \frac{\omega e r^2}{2}. \tag{2.44}$$

But $p_\theta = \mu r^2 \omega$, so

$$M = \frac{e p_\theta}{2\mu} = k \frac{he}{4\pi\mu}. \tag{2.45}$$

It follows that the magnetic moment of a hydrogen or hydrogen-like atom is quantised in units of $he/4\pi\mu$. This basic unit of magnetic moment is called a Bohr magneton, μ_B. It is unfortunate that μ should be the conventional symbol for both reduced mass and magnetic moment, but the use of the subscript B should remove ambiguity.

This prediction of Bohr's theory was checked experimentally by Stern and Gerlach in 1921, who determined the effect of an inhomogeneous magnetic field on a beam of silver atoms. If a small magnet is placed in a homogeneous magnetic field there is no resultant translatory force, but there is in an inhomogeneous field. Consequently, each silver atom experienced a resultant force, determined by its magnetic moment, tending to deviate it from its original path. The atoms passing through the field were allowed to condense on a screen. If the magnetic moment carried by an atom is continuously variable, the effect of the magnetic field should have been to smear the deposit out; but in fact the deposit was separated into discrete sections. The value of the Bohr magneton calculated from this experiment was in good agreement with the theoretical result.

The direction of the magnetic moment is normal to the plane of the orbit, so that a suitable way of defining the external reference axis required for the definition of the third degree of freedom is by applying a magnetic field to the atom. If the angle between the magnetic moment M and the applied field H is ϕ, then the component of the angular momentum along the field direction is p_ϕ. Applying the quantum condition (2.7) to this, we obtain

$$\oint p_\phi d\theta = \cos\phi \oint p_\theta d\theta = m_l h, \quad m_l \text{ integer} \tag{2.46}$$

provided the angle ϕ is constant through the period of the orbit. Recalling (2.21), we have

$$\cos\phi = \frac{m_l}{k}. \tag{2.47}$$

Consequently, not all angles are possible between the orbit normal and the applied field, but only those for which the cosine is an integral multiple of the reciprocal of the angular quantum number. Since $-1 \leqq \cos\phi \leqq 1$, it follows that

$$-k \leqq m_l \leqq k. \tag{2.48}$$

The energy of interaction with the external field H is

$$u = -MH\cos\phi$$

$$= -\frac{eh}{4\pi\mu c}Hm_l \text{ in practical units.} \qquad (2.49)$$

Including this in (2.30), the term series neglecting the fine structure becomes

$$-E_{n,m} = \frac{2\pi e^4\mu}{h^2n^2} + \frac{eh}{4\pi\mu c}m_lH. \qquad (2.50)$$

From which, the spectral wave numbers may be written:

$$\frac{1}{\lambda} = \frac{2\pi^2e^4\mu}{h^3c}\left(\frac{1}{n_1^2} - \frac{1}{n_2^2}\right) + \frac{e}{4\pi\mu c^2}(m_1 - m_2)H \qquad (2.51)$$

where the subscripts 2 and 1 refer to the initial and final states respectively. This means that in the presence of a field, each line of the original spectrum will have a number of additional lines on either side, since $(m_1 - m_2)$ can be positive or negative; and these lines will be evenly spaced, with a spacing proportional to the applied field strength. These facts were discovered experimentally by Zeeman in 1896; and the problem of the separation can be solved on classical grounds, as may be inferred from the absence of Planck's constant h from the separation term in (2.51). Larmor's theorem in classical electrodynamics states that the possible motion of an electron in a magnetic field is exactly the same as the motion in the absence of a magnetic field superimposed on a general rotation about the field direction with a frequency

$$\nu_L = \frac{e}{4\pi\mu c}H \qquad (2.52)$$

called the Larmor frequency. The additional energy is then the "phase integral" $\oint p_\phi d\theta$ multiplied by the Larmor frequency. Inserting the quantum value of the phase integral (2.46) then gives (2.49), showing that the magnitude of the splitting is classical in origin.

The classical analysis of the Zeeman effect is due to Lorentz. In fact, experimentally, each line in the original spectrum splits into three, suggesting a second selection rule:

$$\Delta m_l = \pm 1 \text{ or } 0. \tag{2.53}$$

The Spectra of the Alkali Metals: Spin

In 1890 Paschen and Runge, and completely independently Rydberg, discovered series of lines in the spectra of the alkali metals which were not dissimilar to the series in the spectrum of atomic hydrogen. The chemical monovalence of these metals suggests that one electron is more weakly bound than the rest. The remainder of the atom, comprising the nucleus and all the other orbiting electrons is called the ion core. Plainly there are similarities in the ion core—electron pair and the proton–electron pair. However, the more elliptical orbits of the electron will now penetrate into the ion core, since unlike the proton this is very far from being a point, and thus the electron will "see" a larger positive charge at perihelion. This will "swing" the electron round in a way similar to that of the relativistic mass change, but in the reverse direction, and, one would expect, with a very much larger effect at higher atomic numbers. The result of this is to remove the degeneracy with respect to the quantum number k that was present for the non-relativistic hydrogen atom.

The lines in the principal series in the spectra of the alkalis are doublets, consisting of two closely spaced lines. The best known example of this is the yellow D line of sodium, in which the doublet can be resolved with very simple equipment. The wavelengths of the two components are

$$D_1 = 5896 \text{ Å} \quad \text{and} \quad D_2 = 5890 \text{ Å}$$

so that the transitions producing these lines are between states of very similar energy. In very weak magnetic fields these lines split in a manner quite different to the normal Zeeman effect: both the original lines are deleted, one being replaced by four lines equally spaced about the original line position, and the other by six. The Bohr theory cannot explain this "anomalous Zeeman effect".

Uhlenbeck and Goudsmit in 1925 suggested that since it was not possible to explain this behaviour on the basis of three degrees of freedom, there must be four or more. This is quite reasonable, since an orbital system in which the primary and satellite are not points does have additional internal degrees of freedom. In particular, an electron having physical extent could change its shape, or spin about three axes. If, to take the simplest example, the electron is regarded as a rigid sphere, it can spin about an axis, and thereby possess an internal angular momentum which may be combined with the orbital angular momentum. In fact, it is necessary to postulate only one additional degree of freedom to the electron to explain the alkali doublets, and since a rigid sphere has just one additional degree, it is customary to refer to the internal degree of freedom of the electron as "spin", although it would be altogether wrong to attach too much importance to this. If the spin moment is assumed to be quantised, we may write

$$p_s = s \cdot \frac{h}{2\pi}. \tag{2.54}$$

Similarly, its component along the direction of the angular momentum of the orbit must be quantised. Arguing as before, it would appear that there will be $(2s+1)$ different settings of the spin. In fact, since the lines whose existence we wish to account for are doublets, it appears that

$$s = \tfrac{1}{2} \tag{2.55}$$

and it may be combined with the orbital momentum in only two ways, which are referred to as "spin up" and "spin down".

It is necessary to reanalyse the Bohr theory in the light of the existence of a spin moment, but since in a later chapter the whole problem will be restated in more modern terms, we will neglect this reanalysis here. It is sufficient to observe that the complete specification of a stationary state in the Bohr theory requires the specification of four quantum numbers, three associated with the translational degrees of freedom of the electron, and one, which may have only the values $\pm\frac{1}{2}$, with spin.

The Pauli Exclusion Principle

From the classical point of view all the electrons in a many-electron atom will enter the ground state. There are, however, good reasons for supposing that this is not the case. Since the chemical properties of the elements are determined by the relative ease with which electrons may be transferred from or added to the atom, one would expect a gradual change in properties with atomic number. In fact, as was demonstrated by Mendeleev, the properties of the elements change in a periodic manner as the atomic number increases. In particular, the chemical properties of the elements, and their ionisation potentials, suggest that a small number of their electrons are very much more weakly bound to the atom than the remainder, whereas the classical picture would suggest that all the electrons are equivalent.

If heavy atoms are bombarded with energetic electrons, high energy photons are emitted. These X-ray photons were studied by Mosely who showed in 1913 that the wave numbers were given by an expression similar to that of Rydberg:

$$\frac{1}{\lambda} = R(Z-\delta)^2 k \tag{2.56}$$

where R is the appropriate Rydberg constant, Z is the atomic number, and δ and k are constants that are the same for all the elements, but differ for each line in the X-ray spectrum. In accord with the theory of Bohr, these results can be explained by attributing the X-ray emission to the rejection of a tightly bound electron from the atom by collision with the bombarding electron, followed by an electron from a less tightly bound state falling into the vacant state, with the emission of a quantum. The

X-ray spectrum is characteristic of the element, and is unaffected by chemical binding, confirming that the electrons taking part in the X-ray emission process are not involved in chemical combination. Since the emission of a quantum indicates that an electron has fallen into the tightly bound state, but cannot do so until an electron has been knocked out, it follows that there is a restriction on the number of electrons that can occupy a given state. Analysis of the periodicity of the Periodic Table and consideration of Mosley's law (2.56) leads to the statement: "A quantum state, defined by a specific set of four quantum numbers, can accommodate only one electron at most."

This very important statement is called the Pauli exclusion principle. Its consequences will be dealt with in more detail in a later chapter.

Conclusions

The success of the old quantum theory in accounting for the features of the spectrum of atomic hydrogen, for the qualitative features of the spectra of the heavier atoms, for the Periodic Table, and for a large number of special experimental effects, very strongly suggests that the general approach is good. However, in some important details the theory breaks down, and in particular it requires a quite remarkable collection of supplementary hypotheses and restrictions. Finally, and, as will appear, fundamentally, there is the defect that although the quantum restrictions are introduced, the remainder of the problem is still handled with Newtonian mechanics. The dual nature of the electron and the photon does not appear in the Bohr theory at all. Since it has been demonstrated that this property is fundamental, it is scarcely surprising that the theory fails: the astonishing thing is that it works so well.

CHAPTER 3

The Uncertainty Principle of Heisenberg

The Consequences of Wave–Particle Duality

In Chapter 1 it was demonstrated that duality appeared to be a basic property of both electromagnetic waves and material particles. Since it is plain that an *ad hoc* theory of the type of the old quantum theory will not produce this result, it is worth investigating the theoretical consequences of duality as a preliminary to constructing a mechanics which will contain this fact as a basic property.

Experimentally, a single quantum may be isolated by permitting a beam of quanta to fall on an opaque shutter at a sufficiently low intensity so that the arrival of a quantum at the shutter is a rare event. The shutter is then opened for a short length of time, and closed. In a number of cases, performance of this experiment will admit just one quantum through the shutter: in some cases, of course, more than one quantum, and in others no quantum, will pass through. It will be assumed that a subsequent experiment will eventually be performed, and all cases in which have not resulted in the isolation of a single quantum will be rejected. Suppose further that the incident beam of quanta has passed through a highly accurate collimation and monochromating system, so that in the wave representation a unidirectional monochromatic wave falls on the shutter.

The results of this experiment may be described in either particle or wave terms; and it is a necessary consequence of the existence of duality that these descriptions be completely equivalent, and that in particular neither description should lead to an experimental method of discriminating between the two views, since it is known that in fact none of the experiments performed to date has succeeded in differentiating the particle and wave properties.

From the wave point of view the continuity of the carrier medium must be preserved. This means that when the shutter is opened the wave does not immediately propagate through the opening with its incident amplitude. Instead, the amplitude of vibration of the medium increases progressively with time as the wave passes, eventually reaching a maximum equal to the amplitude of the incident wave. Similarly, the closing of the

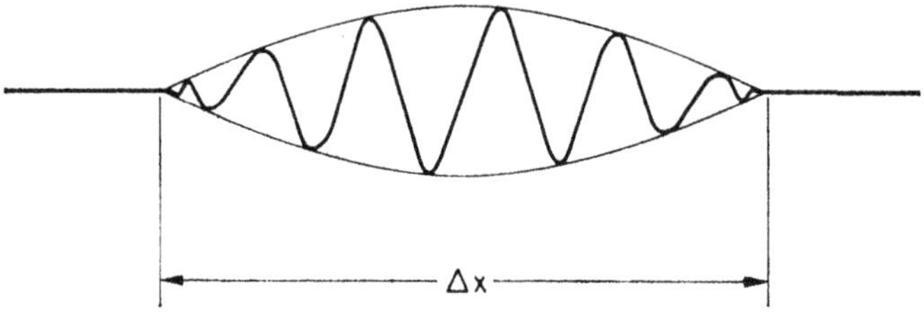

FIG. 3.1. A wave packet.

shutter does not produce an abrupt drop in the amplitude, but a decay towards zero. The resultant shape of the disturbance admitted by the shutter is shown in Fig. 3.1. This type of disturbance is called a "wave group" or "wave packet".

This wave packet is thus the description in wave terms of a single quantum. But plainly the quantity Δx, which is the length of the wave packet, defined as the length outside of which the displacement of the medium is effectively zero, is not in any real sense the "length" of the quantum, since it is determined by the experimental conditions. If the intensity is reduced, the shutter will have to remain open longer in order to admit one quantum. If the propagation velocity of the wave remains unaltered, the packet will then be longer; but if the concept of size has any significance for a quantum then it will be unaltered by such a trivial alteration in the experimental conditions. In addition, it is relatively easy to make the packet much longer than the length of the quantum as measured by other experiments, such as that of Lawrence and Beams on photons.

In the particle model, the quantum is a small particle which passed through the shutter at some time while it was open. If an experiment had previously indicated its position at some earlier time, and its velocity at the same instant, it should then be possible to calculate the position of the particle after passing through the shutter. Suppose that both the previous experiments had contained experimental errors, so that the particle could be located within a distance δx. Then plainly if δx is smaller than Δx the two descriptions will not be equivalent, and a subsequent check on the actual position of the particle will permit a choice between the two descriptions. Since this is contrary to the requirements of the theory, we conclude that δx is not smaller than Δx. Consequently, the length of the wave packet is related to the uncertainty in our knowledge of the position of the quantum. But since as before we can arrange to have Δx any length at all, and since the operations which alter the length of the wave packet cannot affect a previous determination of the position of the quantum, it follows that in fact $\delta x = \infty$, that is, our experiment apparently excludes any previous knowledge of the position of the quantum, and is itself a position measuring experiment. As such, it can determine the position of the quantum without a lower limit on the accuracy, at least in principle, since by increasing the intensity of the incident beam and decreasing the length of time the shutter remains open, Δx may be reduced without limit.

The incident beam is monochromatic, and since de Broglie's equation (1.10) relates wavelength to momentum, it follows that from the particle point of view the incident beam consists of particles all having the same precisely defined momentum. The monochromating system is in fact a momentum measuring experiment in particle terms. However, the wave packet is plainly not a simple monochromatic wave. In fact, a wave group can be represented as a superimposition of a very large number of sine waves with wavelengths in a range about a mean. The amplitudes of each of the components decreases with increasing difference from the mean. The appearance of this "band" of wavelengths in place of the original pure wave can be easily understood in terms of distortion: the operation of the shutter introduces a distortion into the signal, and this distortion produces the group. The need for a band of wavelengths to produce pulses is well known in radar techniques, and also in radio: amplitude modulation high fidelity radio is impossible because of the very large band width necessary to carry the signal.

Consider the very simple case of the superimposition of two sine waves:

$$y_1 = \sin(k_1 x - \omega_1 t); \qquad y_2 = \sin(k_2 x - \omega_2 t).$$

The resultant function has the form

$$Y = y_1 + y_2 = 2\sin\left(\frac{k_1+k_2}{2}x - \frac{\omega_1+\omega_2}{2}t\right) \cos\left(\frac{k_1-k_2}{2}x - \frac{\omega_1-\omega_2}{2}t\right). \quad (3.1)$$

This function is shown graphically in Fig. 3.2, and represents a train of wave groups length Δx: since Y vanishes at $t = 0$ for $x = 0$ and for $x = 2\pi/k_1 - k_2$,

$$\Delta x = \frac{2\pi}{k_1 - k_2} = \frac{2\pi}{\Delta k}$$

or,

$$\Delta x \Delta k = 2\pi. \quad (3.2)$$

The exact form of (3.2) depends on the detailed shape of the wave group, but in general there is always a constant product of the length of a pulse signal and the range of wave numbers required to carry it. In this expression we have introduced the "wave number" k defined by

$$k = 2\pi/\lambda \quad \text{where } \lambda \text{ is the wavelength.}$$

From the wave point of view therefore our experiment to isolate a single quantum has distorted the original pure signal, and the degree of distortion increases as the packet is shortened by reducing the opening time of the shutter.

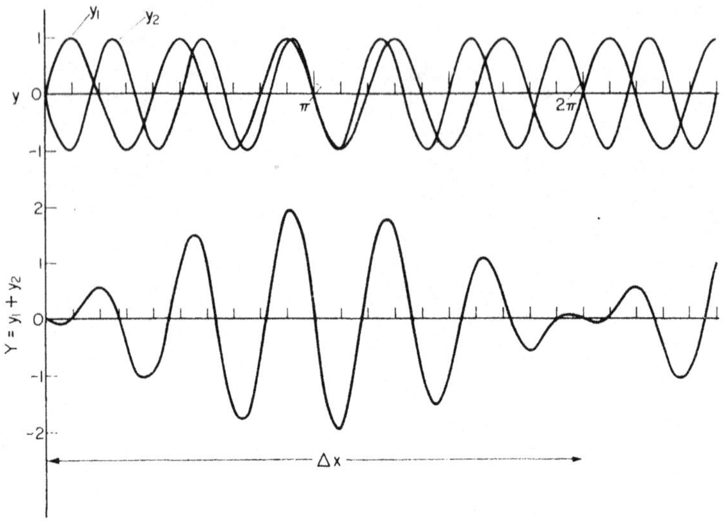

FIG. 3.2.

Now, recalling the de Broglie relation (1.10),

$$p = \frac{hk}{2\pi}$$

$$\Delta k = \frac{2\pi}{h}\Delta p.$$

Substituting in (3.2),

$$\Delta x \Delta p = h. \tag{3.3}$$

What then is the significance of the quantity Δp? It is not possible for a single particle to possess a range of momenta:

if consequently the momentum was known any more exactly than a range Δp, it would be possible to differentiate between the wave and particle views. Our original requirement therefore identifies Δp as an uncertainty in the momentum. But the experimental arrangement was chosen so that the original momentum of the particles was known exactly: it is easy to envisage the appearance of a distortion in the wave signal, but less easy to picture the destruction of knowledge in the particle model. However, it is reasonable to suppose that if the particle is in the immediate vicinity of the shutter when it opens or closes its momentum will be modified by the shutter action. Since we do not know whether the particle was in fact in such a position during the isolation experiment it is clear that the subsequent momentum is to that extent uncertain. The longer the shutter is held open to allow through one particle, the smaller the chance of it interacting with the shutter, and so the smaller the uncertainty in the momentum. This is exactly the same result as the wave model, in which the longer the shutter is held open the less the distortion of the wave. However, this condition is exactly that which leads to a larger uncertainty in the position. Consequently, although there is no limit in principle on the accuracy with which we may know the position or momentum of the quantum separately, our simultaneous knowledge of these two quantities is limited by an expression like (3.3). In detail, the value of the minimum uncertainty depends on the analysis: careful consideration produces a value of $h/4\pi$, but from the present point of view all that is important is the general form of the expression. It should be noted that in the position determining shutter experiment described above, the initial conditions are chosen so that the momentum is known exactly, and it was consequently proved that the uncertainty in the position of any of the quanta is infinite. The experiment reduces the uncertainty in the position, but at the expense of increasing the uncertainty in the momentum. This general principle, indicating as a necessary consequence of accepting duality as a basic fact of nature the existence of an absolute limit of our simultaneous knowledge of the two quantities position and momentum, is called the Uncertainty Principle of Heisenberg, and it was first enunciated in 1925.

Uncertainty relations like (3.3) are formed between any pair of quantities which may be used to set up Newtonian equations of motion. Such pairs of variables are called "canonically conjugate" variables: energy and time being an example.

Bohr's Microscope Experiment

The demonstration that the Uncertainty Principle was a necessary consequence of duality was pointed out by Bohr in 1928, who demonstrated the point most elegantly with a hypothetical experiment in which the position of a particle whose initial momentum is known exactly is determined with a light microscope. Suppose that we have a stream of particles whose momentum is exactly known, or, in wave terms, an accurately monochromated beam, exactly collimated. The direction of propagation is chosen as the x-axis. We now determine the position of one of these particles with a light microscope, with an objective of angular aperture ε, illuminating the particle for the purposes of the determination with a monochromatic and collimated beam of light, wavelength λ. The experimental assembly is shown schematically in Fig. 3.3.

The accuracy with which the position of the particle can be measured with this assembly is limited by the resolving power of the objective. The image of a point produced by a microscope consists not of a point, but of a diffraction pattern, which is composed of a central peak surrounded by concentric rings of alternate maxima and minima. If two particles are sufficiently close together that the centre of the diffraction pattern due to one of them falls within the circle corresponding to the first minimum of the diffraction pattern due to the other, it will be impossible to separate the two. This distance is thus the smallest that can be resolved by the microscope, and we identify this as the uncertainty in the position measurement. The wave theory of light shows that

$$\Delta x \cong \frac{\lambda}{\sin \varepsilon} \tag{3.4}$$

neglecting the refractive index in the object space. Again, this

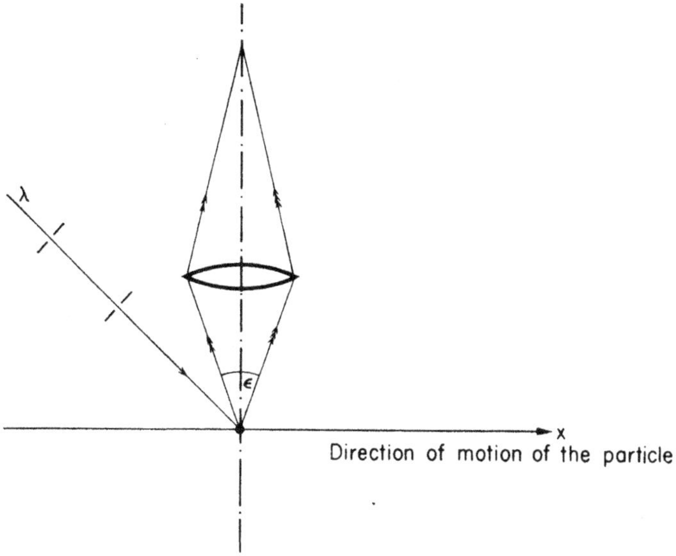

FIG. 3.3. Bohr's microscope.

uncertainty can be reduced without limit in principle by increasing the aperture of the lens and decreasing the wavelength of the illumination, and both methods are in fact used to improve the resolving power of microscopes. However, in order for the particle to be observed at all, it is necessary for *at least* one photon to be scattered from the particle through the optical system. In a scattering process like this both the photon and the particle whose position is being measured will behave as particles, with a resultant change in the momentum of the particle. This phenomenon is well known experimentally, and is called Compton recoil: independent experiments have demonstrated that the momentum change in the collision is of the order of h/λ. If the complete geometry of the collision were known, this momentum change would not introduce any uncertainty in the momentum. However, in order to know the full geometry it is necessary to know one of the resultant paths of the colliding particles (since both the original paths are known exactly, because of the collimating systems). Since the path change of the photon is greatest, it is plain that this will be the easiest to measure. If a path measuring experiment is introduced it will necessarily interact with the photon, thus destroying our knowledge of the position of the particle. The only path measuring experiment is therefore the microscope itself, but after the photon has been focused it is no longer possible to determine which path in the bundle of rays entering the objective it has travelled along. The result is that there is an uncertainty in the post-collision path of the photon which increases with the angular aperture of the objective, and a consequent uncertainty in the momentum of the particle after the collision with the photon of magnitude

$$\Delta p_x \cong \frac{h}{\lambda} \sin \varepsilon \tag{3.5}$$

where Δp_x is the uncertainty in the momentum along the x-direction. Again, this quantity can be reduced without limit, either by increasing the wavelength of the illumination and hence decreasing the momentum change as a result of the Compton recoil, or by decreasing the angular aperture of the microscope and so improving it as a path measuring device. It is plain that the conditions for reducing the momentum uncertainty are exactly the reverse of the conditions required to reduce the position uncertainty, and in fact the product

$$\Delta x \Delta p_x \sim h \tag{3.6}$$

does not contain the experimental conditions at all. The result therefore of applying the property of duality to the illuminating photons in this hypothetical experiment has been to produce an equation exactly like (3.3).

The result is quite general: now that the influence of the experiment is explicitly acknowledged, an uncertainty relationship can be deduced for any experimental arrangement designed to provide simultaneous measurements of any two canonically conjugate variables.

The existence of an Uncertainty Principle demonstrates the invalidity of the Newtonian mechanics in the general case, since the Newtonian equations presuppose an exact knowledge of the dynamical variables of a system, and ignore the necessary experiments: quantum mechanics on the other hand must take specific account of the fact that the quantities we are concerned with are the results of experimental determinations, and that discussion of the values of dynamical variables, as opposed to the results of experimental determinations of the dynamical variables, has no meaning.

Phase Velocity and Group Velocity

The velocity of a wave is the velocity at which a specific peak (or other easily identifiable point) on the wave moves. This is then the velocity of a point having a particular phase, and is called the phase velocity. The velocity of sound in a medium as customarily defined is a phase velocity, and is a property of the propagating medium. However, in general since most observations cannot be performed continuously it is often not possible to measure the phase velocity of a wave. For example, a typical experiment may involve the observation of the position of the maximum at two times. However, unless the wave is under continuous observation, we cannot reidentify the original maximum, since every maximum on a wave looks the same. The concept of continuous observation is lost in quantum mechanics, so that in general the phase velocity of a wave is not meaningful. It is possible to mark the wave by superimposing a second signal, producing in fact a wave group. Then the velocity of this can be measured with ease. However, this velocity is totally unrelated to the phase velocity. In the case of a violin string for example the observed motion of the string corresponds to a "standing wave", that is, one with no velocity. The phase velocity of an elastic wave along a violin string is, however, not zero, but is equal to the velocity of sound in the material of the string under the given tension. It can be demonstrated that if two waves of equal wavelength and amplitude, but opposite directions of propagation are superimposed, the resultant disturbance has the form of a standing wave. In this simple case then the group velocity is zero, while the phase velocities are equal and opposite.

The phase velocity v_{ph} of a wave having the form

$$y = A\cos(\omega t - kx)$$

is

$$v_{\text{ph}} = \omega/k = \nu\lambda. \tag{3.7}$$

The group velocity of a combined wave may be easily found by rewriting (3.1) in the form

$$y_1 = A\cos[(\omega_0 - \Delta\omega)t - (k_0 - \Delta k)x]$$

$$y_2 = A\cos[(\omega_0 + \Delta\omega)t - (k_0 + \Delta k)x]$$

$$Y = y_1 + y_2 = 2A\cos(\omega_0 t - k_0 x)\cos(\Delta\omega . t - \Delta k . x). \tag{3.8}$$

This then represents a modulated wave with an average angular velocity ω_0 in the carrier wave

$$\cos(\omega_0 t - k_0 x) \tag{3.9}$$

and a slowly varying amplitude considered as the modulation

$$\cos(\Delta\omega . t - \Delta k . x) \tag{3.10}$$

The phase velocity of the carrier wave is then

$$v_{\text{ph}} = \omega_0/k_0$$

as before; and it can be seen that the velocity of propagation of the modulation is

$$v_g = \Delta\omega/\Delta k. \tag{3.11}$$

In the limit, as $\Delta\omega, \Delta k \rightarrow 0$, this becomes

$$v_g = \frac{\partial \omega}{\partial k}. \tag{3.12}$$

In this simple case, the modulating wave is in fact the envelope of the wave groups, so that v_g is the velocity of the groups. However, (3.12) is quite general for any shape of group, and consequently the group velocity v_g is always given by (3.12). It is easy to see that both its sign and magnitude is unrelated not only to the velocity of the carrier wave, but also to the phase velocities of the component waves (3.1). Since in accord with the principle of duality no experiment can differentiate between the wave and particle predictions, it follows that the particle velocity is equal to the group velocity.

Combining (3.12) with the de Broglie result (1.9), we obtain

$$\omega = \frac{\hbar k^2}{2m} + \text{const. where } \hbar = \frac{h}{2\pi} \tag{3.13}$$

and recalling the Einstein result (1.1),

$$E = \frac{\hbar^2 k^2}{2m} + \text{const.} \tag{3.14}$$

From (1.9), as k approaches zero, the momentum approaches zero, so the constant in (3.14) is plainly the potential energy V:

$$E - V = \frac{\hbar^2 k^2}{2m}. \tag{3.15}$$

If V is independent of ω and k, it follows from (3.7) and (3.13) that the phase velocity of the wave is a function of the wave number k:

$$v_{\text{ph}} = \frac{\hbar}{2m} k + \frac{V}{\hbar k}. \tag{3.16}$$

This means that the waves composing the group have different phase velocities, with the result that the group will change shape with time. This type of motion is called dispersive motion, and can be inferred from the Uncertainty Principle, since if the wave group did not increase its length in time, a subsequent measurement of its position would provide more accurate information than is allowed by the principle. The motion of a group in an elastic medium, on the other hand, is non-dispersive, since the phase velocity of elastic waves is a property of the medium and independent of the wave number. Consequently, a disturbance in an elastic medium propagates without change of shape.

CHAPTER 4

The Foundations of Quantum Mechanics

Wave Mechanics and Matrix Mechanics

Although the concept of quantisation was suggested as early as 1900, and duality in 1905, there was a twenty-year delay before the Uncertainty Principle, which is a direct consequence of duality, was enunciated. This gap was caused by the conceptual difficulty, and by the unwillingness to abandon deterministic methods. Once the intellectual jump had been made however, the advance of quantum theory was rapid; and in 1926 two systems of mechanics were proposed to replace the inadequate Newtonian system. One, due to Heisenberg, is called matrix mechanics: the other, due to Schrödinger, is called wave mechanics. The results of these two systems are the same, but the mathematical apparatus used appears quite different. Dirac has since demonstrated the formal mathematical equivalence of the two systems, but their apparent difference serves to point out again that in quantum mechanics it is only the prediction of the results of experiments that matters: any question which cannot be tested by some experiment, at least in principle, is outside the realm of quantum theory.

In this book, we shall discuss only the wave mechanics of Schrödinger, since the formulation is more accessible to qualitative arguments. The mechanics involves a set of postulates, but fortunately for our purposes only the two simplest of these are necessary.

The First Postulate of Wave Mechanics: The Wave Equation

Associated with a dynamical system is a function Ψ which is a function of space and time:

$$\Psi = \Psi(x, y, z, t). \tag{4.1}$$

This function has the following properties:

(a) it has wave form with respect to the space coordinates;
(b) different Ψ may be superposed;
(c) its magnitude is large where the system is likely to be in space–time, and small elsewhere;
(d) it is a well-behaved function.

The first two properties are determined by our wish to use this function to account for the phenomenon of interference. The third property is also a consequence of the requirement to explain interference, and is required if the function is to have any immediate physical connection with the properties of the system. The fourth property is a mathematical requirement, although shortly the physical necessity for it will be demonstrated. Good behaviour of a function in this context requires that

(a) the function is continuous everywhere;
(b) the first derivative of the function is continuous everywhere;
(c) the function remains finite as any of x, y, z or t approach $\pm$infinity;
(d) the function is not identically zero.

The first two conditions are called the standard continuity conditions; the third is called the standard boundary condition.

Ψ, which is in general a complex function, is called the Wave Function.

The Wave Equation

In the terms of the previous chapters, we equate the wave function Ψ to the de Broglie wave, and consequently a dynamical system located in space can be regarded as a wave packet, where now the "wave" in this sense means the Schrödinger wave. Exactly as is the case with elastic waves, we expect to be able to obtain a differential equation whose solutions will correspond to the possible forms of the wave packets. For elastic waves, the appropriate equation is easy to derive from first principles and is well known: in one dimension it has the form

$$\frac{\partial^2 y}{\partial t^2} = v^2 \frac{\partial^2 y}{\partial x^2} \tag{4.2}$$

where y is the displacement of the medium at a position x and time t. The general solution of this equation is

$$y = F(x-vt)+G(x+vt) \tag{4.3}$$

where F and G are any functions at all. The solution (4.3) then corresponds to a disturbance whose shape is $F(x)$ when $t = 0$ travelling from left to right with velocity v without change of shape, and a disturbance whose shape is $G(x)$ at $t = 0$ moving at the same velocity in the opposite direction, again without changing shape. It is plain that for this case, the conditions of good behaviour listed above must apply, but otherwise there is no restriction on the form of the functions F and G. These functions are defined by the form of the initial disturbance of the elastic medium, and possibly by additional physical limitations on the motion. In a violin string, for example, the string is clamped at either end, so that there is an additional condition that the amplitude of the motion at these two points is at all times zero. Conditions such as these are called boundary conditions.

Although the general solution of (4.2) has the form (4.3), it is usual to express the functions F and G in terms of a simpler set of functions which are then superposed to produce the appropriate shape. Usually, this simpler set of functions is the set of sine waves, and the mathematical method involved is that of Fourier analysis. It is easy to see that in fact superposition is a property of motion in an elastic medium. In the case of a violin string instanced above, the sine wave solutions to the wave equation

(4.2) that also satisfy the boundary conditions form a set of standing waves, whose wavelengths are integral fractions of twice the string length. These are called the normal modes of the string. The normal mode with the longest wavelength is called the fundamental, and the remainder are called the harmonics. Then since these functions are solutions of the equation satisfying the boundary conditions, any linear combination of them is also a solution of the equation satisfying the boundary conditions. It is a little more difficult to demonstrate that these solutions form a complete set, so that *any* function satisfying the equation and the boundary conditions can be expressed as a linear combination of the normal modes, but this demonstration is possible.

The first property of the wave function ought therefore to be restated as: the wave function is expressible in terms of a set of functions having wave form. *Any* set of functions having wave form and comprising a complete set (in the sense described above) will do. We look therefore for a set of functions having any one of the forms

$$\sin(kx-\omega t);\quad \cos(kx-\omega t);\quad \exp i(kx-\omega t);\quad \exp -i(kx-\omega t). \tag{4.4}$$

The form of (4.3) demonstrates that the propagation of a disturbance along an elastic medium is non-dispersive. However, it has previously been demonstrated that the propagation of a de Broglie wave packet is dispersive (3.17) so that it is plain that (4.2) does not have the correct form for the wave function. Equation (3.14) implies that if the equation is second order with respect to space, as is in fact required by the necessity for periodic solutions, then it must be first order with respect to time. In addition, since the general solution must include wave groups, composed of waves each corresponding to different values of the dynamical variables, it is plain that the dynamical variables must not appear in the coefficients of the equation. We write down a trial equation:

$$\frac{\partial^2\Psi}{\partial x^2} = a\,\frac{\partial\Psi}{\partial t} \tag{4.5}$$

and insert our trial solutions (4.4) in it. The first two trial solutions do not satisfy (4.5), but either of the other two will, provided an appropriate choice of the coefficient is made. In particular, for the trial solution

$$\Psi = \exp i\,(kx-\omega t) \tag{4.6}$$

the equation has the form

$$\frac{\partial^2\Psi}{\partial x^2}+\frac{4\pi m i}{h}\frac{\partial\Psi}{\partial t}=0. \tag{4.7}$$

Again writing $\hbar = h/2\pi$, and multiplying through by $\hbar^2/2m$, we obtain the more convenient form

$$-\frac{\hbar^2}{2m}\frac{\partial^2\Psi}{\partial x^2}=i\hbar\frac{\partial\Psi}{\partial t}. \tag{4.8}$$

Now inserting (4.6) into (4.8), the left-hand side becomes

$$-\frac{\hbar^2}{2m}\frac{\partial^2\Psi}{\partial x^2}\equiv\frac{k^2\hbar^2}{2m}\Psi \tag{4.9}$$

and the right-hand side becomes

$$i\hbar\frac{\partial\Psi}{\partial t}\equiv+\omega\hbar\Psi. \tag{4.10}$$

Recalling the Einstein relation (1.1), and (3.16), equations (4.9) and (4.10) may be written

$$-\frac{\hbar^2}{2m}\frac{\partial^2\Psi}{\partial x^2}+V\Psi=E\Psi=i\hbar\frac{\partial\Psi}{\partial t}. \tag{4.11}$$

This equation now differs from (4.8) because the potential energy V has been explicitly included, using (3.16). The potential energy V may be a function of space and time:

$$V = V(x, y, z, t). \tag{4.12}$$

Equation (4.11) can be further simplified in some circumstances. If the wave function can be expressed as the product of two functions, one of which is a function of position only, and one of time only,

$$\Psi(x, t) = \psi(x)u(t) \tag{4.13}$$

Equation 4.11 becomes

$$-\frac{\hbar^2}{2m}\frac{\partial^2\psi}{\partial x^2}u + Vu\psi = Eu\psi = i\hbar\psi\frac{\partial u}{\partial t}. \tag{4.14}$$

Dividing by $u\psi$,

$$-\frac{\hbar^2}{2m}\cdot\frac{1}{\psi}\cdot\frac{\partial^2\psi}{\partial x^2} + V = E = i\hbar\frac{1}{u}\frac{\partial u}{\partial t}. \tag{4.15}$$

The last term in (4.15) is a function of time only, and it therefore follows that the total energy E is a function of time only. However, if the potential energy is independent of time, and depends only on position, the first part of (4.15) depends only on position. It follows therefore that in this case the total energy of the system E is independent of both position and time. This means that the energy is conserved, so that in fact the separation (4.13) is valid for potential fields in which energy is conserved. Such potentials are called "conservative fields". For these cases we can write two equations:

$$-\frac{\hbar^2}{2m}\frac{\partial^2\psi}{\partial x^2} + V\psi = E\psi \tag{4.16}$$

and

$$i\hbar\frac{\partial u}{\partial t} = Eu. \tag{4.17}$$

Equation (4.16) is called Schrödinger's time independent equation. Equation (4.17) can be integrated directly:

$$u = \exp\left(-\frac{iEt}{\hbar}\right). \tag{4.18}$$

For dynamical systems in conservative fields there is consequently no need to consider the time dependence explicitly. All that has to be done is that (4.16) has to be solved for ψ, then the complete solution of the wave equation can be written down using (4.18) and (4.13).

Eigenfunctions and Eigenvalues

Equations (4.16) and (4.17) are examples of a wide class of equations which can be represented symbolically as

$$\text{operator (function)} = \text{constant.function} \tag{4.19}$$

that is, an operator operating on a function produces an answer which is a constant multiplied by the same function. Equations of this type are called eigenvalue equations. For a given operator, there will be only a limited set of functions for which the equation is satisfied. These functions are called the eigenfunctions of the operator. The corresponding set of constants for which the equation is satisfied are called the eigenvalues of the operator and the complete set is called the eigenvalue spectrum of the operator. An eigenfunction and its corresponding eigenvalue are said to belong to each other.

It does not follow that eigenvalue spectra are necessarily discrete: for example consider the differential operator d/dx. The eigenvalue equation is

$$\frac{dy}{dx} = ay \tag{4.20}$$

the solution of which is

$$y = b \exp(ax). \tag{4.21}$$

where b is an arbitrary constant. This equation is valid for all real and imaginary values of a, including zero. However, if it is also required that the solution be well behaved, it is plain that all real values of a are excluded, since in this case the function becomes infinite as x approaches either $+$ or $-$ infinity. For all imaginary a, however, y has modulus unity and so remains finite as x tends to infinity in either direction. It follows that the eigenvalue spectrum of the operator d/dx is continuous, consisting of all imaginary numbers and zero. Conversely, the operator id/dx has a wholly real eigenvalue spectrum.

Both (4.16) and (4.17) have the form of eigenvalue equations. The energy E may be regarded as the eigenvalue spectrum of the operator

$$-\frac{\hbar^2}{2m}\frac{\partial^2}{\partial x^2}+V \tag{4.22}$$

or the operator

$$i\hbar\frac{\partial}{\partial t}. \tag{4.23}$$

The second of these operators plainly places no restrictions on the possible values of E except that they must be wholly real, as indicated by the discussion of the differential operator above. The first may be more restrictive, since its properties are not immediately obvious, and in fact depend on the functional form of V. This analysis leads to the second Schrödinger postulate:

The Second Postulate of Wave Mechanics

The possible results of an experimental determination of the value of a dynamical variable of a system are the terms of the eigenvalue spectrum of a suitably chosen operator operating on the wave function.

This postulate specifically takes account of the fact that we are dealing with the results of experiments: it consequently does not lead to a statement like "the velocity of a particle is . . ." but merely "the result of an experimental determination of the velocity may be . . .". Even in terms of a classical model these two statements plainly differ, if one takes account of the structure of the experiment. Suppose, for example, one measures the position of a ball on a table by pressing down a grid on to the table and noting which part of the grid the ball is in. This experiscent can never give the result that the ball is in a position corresponding to the bars of the grid, since as the grid comes down the ball will be pressed to one side into one of the openings. This is a clear example of an experimental assembly which will only produce discrete results from an initially continuous situation. In the realm for which quantum mechanics is valid the effects of all experiments are coarse in comparison to the quantities measured, so that all experiments to some extent predicate their results. This is implied by the Uncertainty Principle, and is expressly included in the formulation of the Second Postulate.

The question answered by the second postulate does not, of course, exist in the classical problem, for which an ideal non-interacting experiment can always be assumed. Consequently, there is as yet no contact with the laws of classical mechanics, and the questions which classical mechanics is designed to answer—for example, the values of the dynamical variables at some time t after the initial values at $t = 0$ have been specified—cannot be answered in terms of the first two postulates. It is necessary to solve the appropriate Schrödinger equation, to build up the wave packet corresponding to the system, analyse its subsequent movement, and consequent dispersion, and finally operate on the new set of wave functions comprising the packet to obtain the most probable result of an experimental determination of the dynamical variables at that time. Because of the dispersion in the propagation of the packet, the answers are always in terms of the "most probable result" for this type of problem. The solution of this problem requires additional postulates, and the mathematics becomes more complicated. Fortunately, we need not initially consider these problems. It is sufficient for most discussions to know the possible values of the dynamical variables: later other techniques will be developed to derive actual numerical results.

In order to determine the spectra of possible experimental values of the dynamical variables it is necessary to choose the appropriate operators associated with these dynamical variables. Again, this involves a postulate, this time based on the Heisenberg Uncertainty Principle; but again, fortunately, this problem is not important in elementary solid state theory. For our present purposes it is enough to know the spectrum of possible energies of the system, so that we only need to know the operator associated with energy. This is, however, simply the operator (4.22): so that our problem reduces to determining the eigenvalue spectrum of the operator (4.22); or to put it in another way, to solving the time independent Schrödinger equation (4.16). The operator (4.22) is often called the Hamiltonian operator H for the system, and the Schrödinger equation (4.16) written

$$H\psi = E\psi. \tag{4.24}$$

The problem we are going to solve is the determination of the possible values of the experimental measurement of the energy of the dynamical system in a potential field V. The set of possible values we call the energy spectrum, and we solve the problem by writing down the Hamiltonian of the system, by inserting the appropriate form of V in the general equation (4.22), and then determine its eigenvalue spectrum. For our purposes consequently the eigenfunctions are not of primary importance.

Generalisation to Three Dimensions

The Schrödinger equation (4.16) and the Hamiltonian (4.22) have been written down for the one-dimensional case, where the only space variable is x. In this case, the kinetic energy T is given by

$$T = \frac{p_x^2}{2m}. \tag{4.25}$$

In the three-dimensional case, this becomes

$$T = \frac{|p|^2}{2m} = \frac{1}{2m}(p_x^2+p_y^2+p_z^2). \tag{4.26}$$

The operator associated with the kinetic energy in the one-dimensional case is

$$-\frac{\hbar^2}{2m}\frac{\partial^2}{\partial x^2}(\equiv E-V). \tag{4.27}$$

Comparing (4.25), (4.26) and (4.27), it appears that the operator associated with the kinetic energy in the three-dimensional case should be

$$-\frac{\hbar^2}{2m}\left(\frac{\partial^2}{\partial x^2}+\frac{\partial^2}{\partial y^2}+\frac{\partial^2}{\partial z^2}\right). \tag{4.28}$$

The operator

$$\left(\frac{\partial^2}{\partial x^2}+\frac{\partial^2}{\partial y^2}+\frac{\partial^2}{\partial z^2}\right)$$

is called the Laplacian operator, written ∇^2, so that in three dimensions the time independent Schrödinger equation is written

$$-\frac{\hbar^2}{2m}\nabla^2\psi+V(x,y,z)\psi = E\psi. \tag{4.29}$$

In polar coordinates (r, θ, ϕ) defined by the relations

$$\begin{aligned} x &= r\sin\theta\sin\phi \\ y &= r\sin\theta\cos\phi \\ z &= r\cos\theta \end{aligned} \tag{4.30}$$

the Laplacian operator becomes

$$\frac{\partial^2}{\partial r^2}+\frac{2}{r}\frac{\partial}{\partial r}+\frac{1}{r^2\sin\theta}\frac{\partial}{\partial\theta}\left(\sin\theta\frac{\partial}{\partial\theta}\right)+\frac{1}{r^2\,\mathrm{sjn}^2\,\theta}\frac{\partial^2}{\partial\phi^2}. \tag{4.31}$$

This form of the Laplacian is more convenient for problems in which the potential has spherical symmetry.

The Significance of the Wave Function

As shown in the previous sections, a consideration of the physical significance of the wave function is not of primary importance in solving the fairly simple problem of the form of the energy spectrum for a dynamical system in a potential field. However, in the majority of cases it is important to determine the eigenfunctions of the Hamiltonian, and for qualitative arguments it is helpful to have a physical model of the wave function. Furthermore, it is meaningful to consider whether the wave function Ψ is related to displacements in a field, such as is the case with the similar wave associated with photons in the electromagnetic field, or displacement in a medium, as for sound waves, or something different again.

Born suggested that the wave function should be interpreted as a probability function, and in particular suggested that the probability $P dxdydzdt$ that the system occupied a position (x, y, z, t) within a range $dxdydzdt$ was related to the amplitude of the wave function at that position by the equation

$$P(x, y, z, t)dxdydzdt = \Psi\Psi^* dxdydzdt \tag{4.32}$$

where Ψ^* is the complex conjugate of Ψ.

If (4.32) is integrated over all space and time, then since the system must be somewhere at some time, the integral must equal unity:

$$\int_{-\infty}^{+\infty}\int_{-\infty}^{+\infty}\int_{-\infty}^{+\infty}\int_{-\infty}^{+\infty} \Psi\Psi^* dxdydzdt = 1. \tag{4.33}$$

The magnitude of Ψ is always adjusted to satisfy (4.33). This process is called "normalisation", and this interpretation of the wave function is called the statistical interpretation.

In terms of this interpretation it is possible to understand the conditions for good behaviour of the wave function. Plainly, unless the standard boundary condition holds, and unless the function is not identically zero, (4.33) cannot be satisfied. There is no physical meaning in a discontinuously changing probability function, and plainly also the gradient must vary continuously, or the different "probability flow" away from the discontinuity in slope will produce a discontinuity in the probability.

CHAPTER 5

Some Problems in Wave Mechanics

A Free Particle Moving in One Dimension

Consider a free particle mass m travelling in one dimension. Since the particle is free the potential has the form

$$V(x) = 0. \tag{5.1}$$

Inserting this potential, the one-dimensional Hamiltonian operator of the system (4.22) becomes

$$-\frac{\hbar^2}{2m}\frac{\partial^2}{\partial x^2}. \tag{5.2}$$

Consequently, neglecting the constant term $\hbar^2/2m$, the problem reduces to determining the eigenvalue spectrum of the operator $-d^2/dx^2$. The eigenvalue equation (4.19) becomes

$$\frac{d^2}{dx^2}y + \lambda y = 0 \tag{5.3}$$

where y is the eigenfunction and λ the eigenvalue. Setting $|\lambda| = a^2$, there are three possible forms of the solution depending on the sign of λ.

(i) λ positive:

$$y = A \sin ax + B \cos ax, \tag{5.4}$$

(ii) λ negative:

$$y = C \exp(ax) + D \exp(-ax), \tag{5.5}$$

(iii) λ zero:

$$y = Fx + G, \tag{5.6}$$

where A, B, C, D, F and G are arbitrary constants. The solution (5.4) is well behaved for all A, B since both functions remain finite as x approaches infinity in either direction and both the function and its derivative are everywhere continuous. The solution (5.6) is also well behaved provided $F = 0$, and G is non-zero. However, for all non-zero C and D the solution (5.5) approaches infinity at either + or − infinity, and if both C and D are zero the solution is identically zero. Consequently, the eigenvalue spectrum of the operator $-d^2/dx^2$ belonging to well-behaved eigenfunctions consists of all positive numbers and zero. Applying this result to the operator (5.2), we conclude that the energy of a free particle may have any positive value or zero. Consequently, the energy of a free particle is continuously variable but always positive. This is exactly the classical assumption of the possible energy of a free system, so that it may be concluded that the motion of a free particle will be adequately described by Newtonian mechanics, as is in fact the case: the motion of electrons in a vacuum can be described in Newtonian terms.

A Particle Travelling in One Dimension Incident on a Potential Barrier

Now consider the case of a particle mass m travelling from left to right in a potential field which has a sharp discontinuity, height V, at $x = 0$. The potential V has the form

$$V(x) = 0, \qquad x < 0;$$

$$V(x) = V_1, \qquad x > 0 \tag{5.7}$$

and is sketched in Fig. 5.1. For $x < 0$, the Hamiltonian has the same form as (5.2), and consequently the Schrödinger equation (4.24) may be written

$$-\frac{\hbar^2}{2m}\frac{\partial^2\psi}{\partial x^2} = E\psi. \tag{5.8}$$

The general solution may be written

$$\psi_1 = A \exp(ik_1x) + B \exp(-ik_1x) \tag{5.9}$$

where

$$k_1 = \sqrt{\left(\frac{2mE}{\hbar^2}\right)}. \tag{5.10}$$

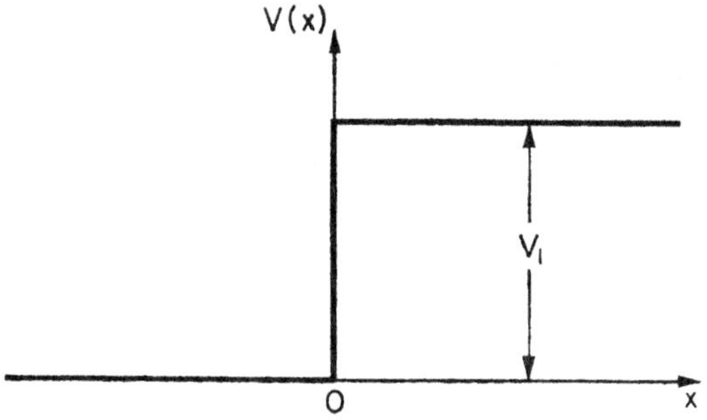

FIG. 5.1. The potential step (5.7).

For $x > 0$, the Hamiltonian has the form

$$H_2 = -\frac{\hbar^2}{2m}\frac{\partial^2}{\partial x^2} + V_1 \tag{5.11}$$

and the Schrödinger equation becomes

$$-\frac{\hbar^2}{2m}\frac{\partial^2\psi}{\partial x^2} = (E-V)\psi. \tag{5.12}$$

The form of the solution depends on the magnitude of the energy of the incident particle. If $E > V$, the solution is similar to (5.9):

$$\psi_2 = C \exp(ik_2x) + D \exp(-ik_2x) \tag{5.13}$$

where now

$$k_2 = \sqrt{\left[\frac{2m(E-V)}{\hbar^2}\right]}. \tag{5.14}$$

Including the time factor (4.18), the first term in both (5.9) and (5.13) represents a wave travelling from left to right, and the second term in both equations represents a wave travelling from right to left. The existence of a wave travelling from right to left in (5.13) implies a probability source to the right of the barrier, but since the particle is incident on the barrier from the left, there can be no such source. Hence

$$D = 0. \tag{5.15}$$

The standard continuity conditions at the barrier require that

$$\begin{aligned} \psi_1(0) &= \psi_2(0) \\ \psi_1'(0) &= \psi_2'(0) \end{aligned} \tag{5.16}$$

where the prime indicates differentiation with respect to x. Applying these conditions to (5.9) and (5.13) we obtain

$$\begin{aligned} A+B &= C \\ k_1(A-B) &= k_2 C. \end{aligned} \tag{5.17}$$

Hence,

$$\begin{aligned} A &= \frac{k_1+k_2}{k_1-k_2} B \\ A &= \frac{k_1+k_2}{2k_1} C \end{aligned} \tag{5.18}$$

The first term in (5.9) represents the incident probability wave, the second represents the reflected probability wave, and the remaining term in (5.13) represents the transmitted probability wave. Recalling the normalisation condition (4.33), and remembering that the experimental arrangement requires that the particle is *certainly* incident on the barrier, we set

$$|A|^2 = 1. \tag{5.19}$$

The amplitude of the reflected probability wave is then

$$\left|\frac{B}{A}\right|^2 = \frac{(k_1-k_2)^2}{(k_1+k_2)^2} \tag{5.20}$$

and the amplitude of the transmitted probability wave is

$$\left|\frac{C}{A}\right|^2 = \frac{4k_1^2}{(k_1+k_2)^2}. \tag{5.21}$$

At first sight, the principle of conservation would appear to require that these amplitudes sum to unity, since the particle is certainly either reflected or transmitted; but since the wave number k is different in the two regions the velocity of the wave is also different. Consequently, it is the reflected and transmitted probability *fluxes* that should sum to unity. The group velocity in the region to the left of the barrier is

$$v_1 = \frac{\hbar}{m} k_1 \tag{5.22}$$

and to the right of the barrier this becomes

$$v_2 = \frac{\hbar}{m} k_2 \tag{5.23}$$

The relative velocity on the right of the barrier is thus k_2/k_1, and the ratio of the transmitted probability flux to the incident flux is

$$\frac{k_2}{k_1}\left|\frac{C}{A}\right|^2 = \frac{4k_1k_2}{(k_1+k_2)^2}. \tag{5.24}$$

The sum of the reflected and transmitted fluxes is then

$$\frac{(k_1-k_2)^2}{(k_1+k_2)^2} + \frac{4k_1k_2}{(k_1+k_2)^2} = 1 \tag{5.25}$$

in agreement with the principle of conservation.

Recalling that in wave optics the refractive index between two media is defined as the ratio of the velocity of light in the two, we may define a refractive index for this type of barrier:

$$\mu = \frac{v_2}{v_1} = \frac{k_2}{k_1} = \sqrt{\left(1-\frac{V}{E}\right)}. \tag{5.26}$$

In terms of the refractive index, (5.20) becomes

$$\left|\frac{B}{A}\right|^2 = \left(\frac{1-\mu}{1+\mu}\right)^2, \tag{5.27}$$

(5.21) becomes

$$\left|\frac{C}{A}\right|^2 = \left(\frac{2}{1+\mu}\right)^2 \tag{5.28}$$

and (5.24) becomes

$$\frac{k_2}{k_1}\left|\frac{C}{A}\right|^2 = \mu\left(\frac{2}{1+\mu}\right)^2. \tag{5.29}$$

It follows that there is always a finite probability that the incident particle will be reflected from the potential barrier (5.7), contrary to the classical prediction that the particle will certainly be transmitted provided $E > V_1$.

Now consider the case where the particle energy is less than the potential step, $E < V$. In this case, the solution of the Schrödinger equation to the right of the barrier becomes

$$\psi_2 = C\exp(a_2x) + D\exp(-a_2x) \tag{5.30}$$

where

$$a_2 = \sqrt{\left[\frac{2m}{\hbar^2}(V-E)\right]}. \tag{5.31}$$

The Standard Boundary Condition gives

$$C = 0 \tag{5.32}$$

otherwise the solution approaches infinity as x approaches plus infinity. Note that the limit as x approaches minus infinity is irrelevant, since (5.30) is only valid in the range 0 to + infinity. The continuity conditions at the step (5.16) give

$$A+B = D$$

$$ik_1(A-B) = a_2D. \tag{5.33}$$

Hence,

$$B = -\frac{a_2 - ik_1}{a_2 + ik_1} A; \qquad D = \frac{2ik_1}{a_2 + ik_1} A. \tag{5.34}$$

Again, the first term in (5.9) represents the incident wave, and the second the reflected wave. The probability that the particle will be reflected is thus

$$\left|\frac{B}{A}\right|^2 = \frac{a_2 - ik_1}{a_2 + ik_1} \cdot \frac{a_2 + ik_1}{a_2 - ik_1} = 1 \tag{5.35}$$

in agreement with the results of classical mechanics, which also predict total reflection. What then is the meaning of (5.30)? It is plain that the amplitude is non-zero for $x > 0$, since the continuity conditions (5.34) require that $D \neq 0$. The probability that the particle will be found at a position x beyond the barrier is

$$\left|\frac{\psi_2(x)}{A}\right|^2 = \frac{4k_1^2}{a_2^2 + k_1^2} \exp(2a_2 x) \tag{5.36}$$

and this quantity is real and finite for all positive x. However, the group velocity in the region to the right of the barrier is

$$v_2 = \frac{i\hbar}{m} a_2 \tag{5.37}$$

so that the probability flux is imaginary. In particular, the refractive index between the two regions, as defined by (5.26), is imaginary. The solution (5.30) might equally well have been written in the form (5.13), but the wave number k_2 would then have been wholly imaginary:

$$k_2 = ia_2. \tag{5.38}$$

The result is that although there is a finite possibility of finding the particle at a position to the right of the barrier, its velocity and momentum are wholly imaginary, and consequently not detectable by an experiment. We interpret the solution to the right of the potential step by saying that unlike the classical result total reflection does not take place at the point $x = 0$. Although the particle is certainly reflected, the position at which this reflection takes place cannot be completely defined. In future, we shall interpret particles with imaginary wave numbers as being in the process of being totally reflected. Since the conclusion of this analysis differs from the classical result, it is worth considering whether an experiment can be designed to test the different predictions.

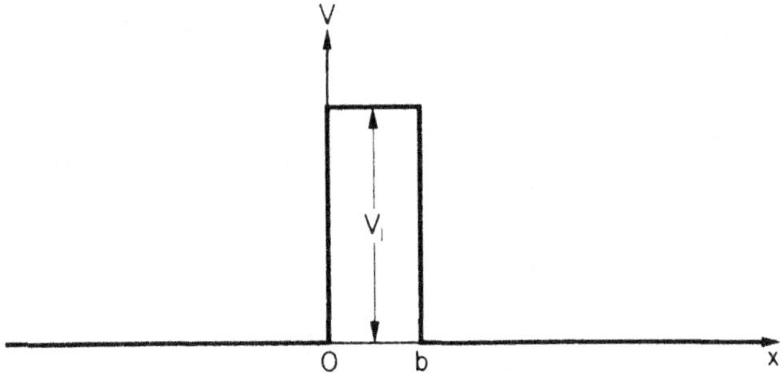

FIG. 5.2. The potential barrier (5.39).

The Tunnel Effect

If the amplitude of the wave function remains finite beyond the potential step, then the introduction of a second step of reverse sign at a position $x = b$ reducing V again to zero should produce a barrier which is classically impenetrable, but does exhibit non-zero transmission in terms of quantum mechanics. Consider a potential of the form

$$V(x) = 0, \quad x < 0 \quad \text{and} \quad x > b$$

$$V(x) = V_1, \quad 0 < x < b. \tag{5.39}$$

This potential is sketched in Fig. 5.2.

Now consider a particle incident on the barrier from the left. The wave function has the form

$$\begin{aligned} x < 0 \qquad & \psi_1 = \exp(ik_1x) + B\exp(-ik_1x) \\ 0 < x < a \qquad & \psi_2 = C\exp(a_2x) + D\exp(-a_2x) \\ x > a \qquad & \psi_3 = F\exp(ik_1x) \end{aligned} \tag{5.40}$$

for $E < V_1$. Equation (5.40) has been simplified in accordance with the methods used before: the incident wave has been normalised, so that in terms of the earlier notation $A = 1$; and the probability wave incident on the barrier from the right has been excluded. k_1 and a_2 have the values (5.10) and (5.31) respectively. It should be noted that although it may be assumed that the amplitude of the wave function diminishes in the barrier, it does not follow that $D = 0$. Applying the standard continuity conditions at $x = 0$ and $x = a$, we obtain

$$\begin{aligned} 1+B &= C+D \\ ik_1(1-B) &= k_2(C-D) \\ C\exp(a_2b)+D\exp(-a_2b) &= F\exp(ik_1b) \\ a_2\{(C\exp(a_2b)-D\exp(-a_2b)\} &= ik_1F\exp(ik_1b). \end{aligned} \tag{5.41}$$

Solving for the constants, and noting that the probability of transmission is $|F|^2$:

$$|F|^2 = \frac{[2(k_1/a_2)]^2}{[1-(k_1^2/a_2^2)]^2\sinh^2 a_2b + [2(k_1/k_2)]^2\cosh^2 a_2b}. \tag{5.42}$$

Consequently, there is always a finite probability of transmission through a barrier of this form. This arises because the effect of decreasing the potential to zero at $x = b$ is to make the wave number real again. But since the wave function cannot have zero amplitude and slope at the end of the barrier, it follows that the real wave number solution on the right of the barrier has finite probability, and hence there is a finite probability of transmission. Since in (5.13) the distance beyond the first step x and the height of the barrier $V-E$ both appear inside the exponential term, it is to be expected that the probability of transmission will fall exponentially with increasing barrier thickness and height.

This strictly non-classical behaviour is called the tunnel effect. Classically, the barrier would be impenetrable for all $V > E$, all b. However, there are a number of experimental manifestations of the tunnel effect, which confirm the fact that the transmission of the barrier falls exponentially with increasing height and thickness.

In optics, a situation analogous to the tunnel effect is well known. If light is shone onto a medium of lower refractive index from a medium of higher, then for angles of incidence greater than a certain value, the wave is totally reflected. Under these conditions, the refractive index between the two media becomes imaginary. If now a second block of the dense medium is brought close to the first, so that between the two there is a thin layer of the less dense medium, the situation is clearly closely analogous to the potential (5.39), since two regions in which the refractive index is real are separated by a thin layer where the refractive index is imaginary. In this situation light is in fact transmitted across the gap, and the transmission of the assembly increases exponentially as the thickness of the barrier layer diminishes.

Aluminium is always covered by a very tenacious layer of aluminium oxide in air. The thickness of this layer is of the order of 20–30 Å. Aluminium oxide is an excellent insulator, so that if two aluminium wires are twisted together they are separated by a thin layer through which electrons cannot pass. In fact, however, it is a matter of everyday experience that current does in fact pass under these circumstances. The electrons penetrate the insulating layer by the tunnel effect.

The emission of particles from radioactive nuclei is also an example of tunnelling, as also is the cold emission of electrons from metals in a high field: finally, the operation of the Esaki tunnel diode depends on the tunnel effect. This last case will be analysed in some detail by Leck in his book *Theory of Semiconductor Junction Devices.*

These experimental proofs of the predicted tunnel effect can be regarded as a justification of the wave mechanical approach.

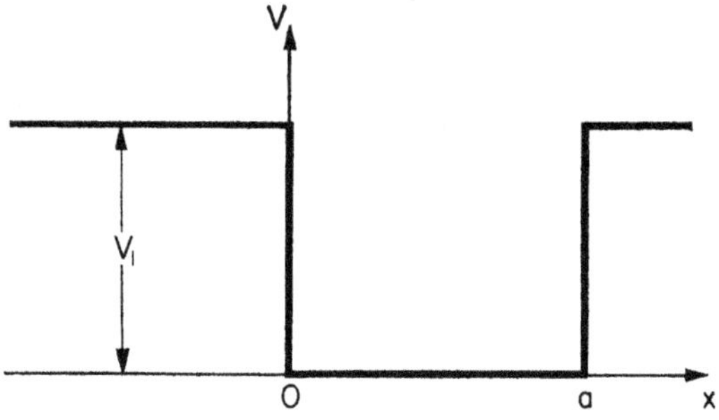

FIG. 5.3. The potential box (5.43).

A Particle in a Potential Box

So far, only situations where the eigenfunctions are unbounded have been considered; that is, the wave function remains finite at either $\pm$ infinity or both. Now an example will be considered for which the wave function approaches zero at both plus and minus infinity. Consider the potential field

$$V(x) = V_1 \qquad x < 0 \quad \text{and} \quad x > a$$

$$V(x) = 0 \qquad 0 < x < a. \tag{5.43}$$

This potential is illustrated in Fig. 5.3. For particles with energy E less than V_1 the wave function may be written

$$\left.\begin{aligned} x < 0, &\quad \psi_1 = A \exp k_1 x \\ 0 < x < a, &\quad \psi_2 = B \sin k_2 x + C \cos k_2 x \\ x > a, &\quad \psi_3 = D \exp -k_1 x \end{aligned}\right\} \tag{5.44}$$

where the standard boundary conditions have been applied to simplify ψ_1 and ψ_3, and ψ_2 has been written in the trigonometric rather than the exponential form. In (5.44)

$$k_1^2 = \frac{2m}{\hbar^2}(V-E); \qquad k_2^2 = \frac{2m}{\hbar^2}E.$$

It is rather tedious to solve (5.44) completely, but a simple form of the problem may be obtained by allowing V_1 to approach infinity. As demonstrated before, the effect of this is to reduce the amplitude of the exponential wave functions ψ_1 and ψ_3 until in the limit

$$\operatorname*{Lim}_{V_1 \to \infty} (\psi_1, \psi_3) = 0. \tag{5.45}$$

The standard continuity condition is then

$$\psi_2(0) = 0; \qquad \psi_2(a) = 0. \tag{5.46}$$

A second continuity condition cannot be written, because in the limit (5.45) the gradient of the functions ψ_1 and ψ_3 at $x = 0$ and $x = a$ becomes indeterminate. The condition (5.46) gives

$$C = 0; \qquad \sin k_2 a = 0 \tag{5.47}$$

and hence

$$k_2 a = n_x \pi \qquad (n_x = 1, 2, 3, \ldots). \tag{5.48}$$

This is analogous to the classical problem of the modes of vibration of an elastic string clamped at either end, and (5.48) represents a set of standing waves whose wavelengths are integral fractions of $2a$. The possible values of the energy E are

$$E = \frac{\hbar^2 k_2^2}{2m} = \frac{\pi^2 \hbar^2}{2ma^2} \cdot n_x^2 \tag{5.49}$$

The eigenvalue spectrum for the Hamiltonian of the system is thus discrete, consisting of an infinite set of "energy levels" numbered by the integer n_x. The number n_x is called a quantum number. The lowest energy in the spectrum is

$$E_1 = \frac{\pi^2 \hbar^2}{2ma^2}. \tag{5.50}$$

The energy cannot have the value zero, for in this case $n_x = 0$ and hence $k_2 = 0$ and ψ_2 is identically zero, contrary to the requirements of good behaviour.

Three-dimensional Potential Box

The derivations for the one-dimensional box may be simply generalised for a three-dimensional case. The potential is

$$\left.\begin{aligned} V(x, y, z) &= 0, \quad 0 < x < a; 0 < y < b; 0 < z < c \\ V(x, y, z) &= \infty, \quad x < 0, x > a; y < 0, y > b; \\ &\qquad z < 0, z > c. \end{aligned}\right\} \tag{5.51}$$

The Hamiltonian for the system is then

$$H = -\frac{\hbar^2}{2m}\left(\frac{\partial^2}{\partial x^2} + \frac{\partial^2}{\partial y^2} + \frac{\partial^2}{\partial z^2}\right) \tag{5.52}$$

which can be separated into three parts, each of the same form as the Hamiltonian for the one-dimensional case. Application of the boundary conditions separately to these three equations gives the conditions

$$\left.\begin{aligned} k_x a &= n_x \pi \\ k_y b &= n_y \pi \\ k_z c &= n_z \pi \end{aligned}\right\} \tag{5.53}$$

where n_x, n_y and n_z are integers. The quantities k_x, k_y and k_z are the three components of the "wave vector" $\mathbf{k}$, whose magnitude is the reciprocal of the wavelength multiplied by 2π, and whose direction is the direction of propagation of the wave. The energy of the states now becomes

$$E = \frac{\hbar^2 |\mathbf{k}|^2}{2m} = \frac{\hbar^2 \pi^2}{2m}\left\{\frac{n_x^2}{a^2} + \frac{n_y^2}{b^2} + \frac{n_z^2}{c^2}\right\} \tag{5.54}$$

where now the set of numbers n_x, n_y and n_z labels the state. These three quantum numbers may also be regarded as the components of a vector **n**, which is called the quantum vector. Setting $a = b = c$, (5.54) becomes

$$E = \frac{\hbar^2 \pi^2}{2ma^2} |\mathbf{n}|^2. \tag{5.55}$$

The vectors **k** and **n** may be graphically represented in a suitable three-dimensional Cartesian reference system as lines joining the origin to the points (k_x, k_y, k_z) or (n_x, n_y, n_z) so that the length of the line is a measure of the magnitude of the vector, and the direction of the line is parallel to the direction of the vector. These three-dimensional "spaces" used to graph the vectors **k** and **n** are called **k**-space and **n**-space respectively. Equations (5.54) and (5.55) show that the energy of a state depends only on the magnitude of the wave or quantum vectors, and not on their direction. Hence the surface of a sphere drawn in either of these two spaces about the origin joins points of the same energy.

Because the solutions of the wave equation have been written in the form of standing waves for mathematical convenience, the integers n are restricted to positive values only. Normally, a change in sign would reverse the direction of propagation, but since the solutions have been written as standing waves this is meaningless. To avoid counting each eigenfunction more than once, therefore, the quantum numbers must be restricted to positive values. Since the n's are restricted to integral values, the permitted values of the vector **n** are represented in **n**-space by a set of lines joining the origin to a discrete set of points in the space, one unit apart. A two-dimensional section through **n**-space is shown in Fig. 5.4, in which the possible end-points of the graphical representations of the vectors **n** are shown by dots. We say that the permitted values of the vector **n** map out a cubic lattice of points in **n**-space. Similarly, a lattice of points is mapped out in **k**-space, and for a cubic box this lattice is also cubic. Each lattice point represents a possible state of the system, so that for **n**-space there is one permitted state per unit volume. In order to facilitate calculation it is convenient to represent this discrete situation by means of a continuous (and therefore integrable) function. The function chosen is called the density of states function, $g(E)$ where $g(E)dE$ is the number of states per unit volume of the potential box ($a^3 = 1$) having energies between E and $E+dE$. The number of states G' having an energy less than a given value E is equal to $\frac{1}{8}$ of the volume of the sphere whose surface is the constant energy surface E in

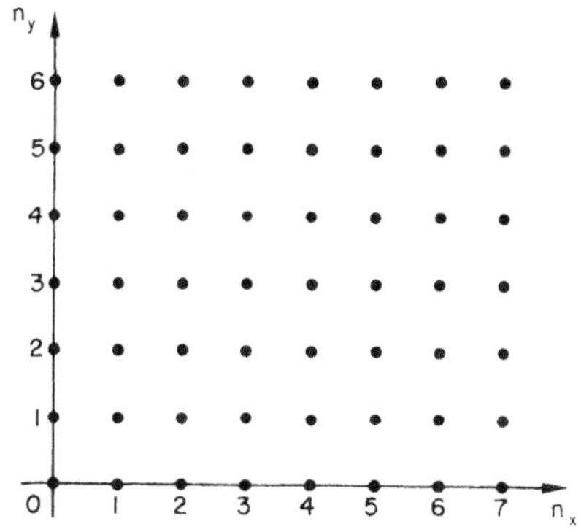

FIG. 5.4. A two-dimensional section through **n**-space: standing wave solutions.

n-space, since there is one state per unit volume of **n**-space. The term $\frac{1}{8}$ appears because the n are restricted to the positive octant only. The radius of this sphere is

$$n_E = \frac{a}{\pi}\sqrt{\left(\frac{2mE}{\hbar^2}\right)} \tag{5.56}$$

and its volume is

$$\tfrac{4}{3}\pi n_E^3 = \frac{4a^3}{3\pi^2}\left(\frac{2mE}{\hbar^2}\right)^{\frac{3}{2}}. \tag{5.57}$$

The number of states per unit volume of the potential box

having an energy between 0 and E is $G'/a^3 = G$, and in terms of the density of states function,

$$G = \int_0^E g(E)dE = \frac{1}{6\pi^2}\left(\frac{2mE}{\hbar^2}\right)^{\frac{3}{2}}. \tag{5.58}$$

Differentiating with respect to E,

$$g(E) = \frac{1}{4\pi^2}\left(\frac{2m}{\hbar^2}\right)^{\frac{3}{2}} E^{\frac{1}{2}}. \tag{5.59}$$

This parabolic function is then the density of states function for a single particle in a potential box with perfectly reflecting sides.

It is possible to write the solutions of the Hamiltonian (5.52) in the form of running waves instead of standing waves. This description is more convenient for problems involving transport of the particle, and in this case the lattice of points in **k**-space is not confined to the positive octant. However, the lattice is 8 times less dense, so the density of states function is unaffected by the change in the description.

When the potential outside the box is not infinite, the wave function does not vanish at the walls of the box. The walls are then no longer perfectly reflecting. The effect of this is to modify the detailed form of the expressions obtained, but provided $E \ll V$, the changes are insignificant in effect.

Many-particle Systems and the Pauli Exclusion Principle

The wave-mechanical many-body problem is just as insoluble as the classical many-body problem. In both cases, extreme approximations are required to produce solutions to even simple problems. There is, however, one important property of the solutions to many-body problems which can be derived without considering how the solutions are obtained. Consider a system of two identical particles, labelled 1 and 2. An observation of the system gives the result that particle 1 is at a position with coordinates a, while particle 2 is at a position with coordinates b. The wave function for the system can be written $\Psi(a, b)$ where this is the total time dependent wave function, including spin, if any. Suppose that an operation is performed on the system which has the effect of reversing the positions of the two particles, so that particle 1 is now at position b and particle 2 at position a. The wave function is now $\Psi(b, a)$. Because the two particles are identical, it will not be possible experimentally to distinguish this situation from the original, and therefore any operator operating on the wave function $\Psi(a, b)$ will produce the same result as the same operator operating on the wave function $\Psi(b, a)$. Consequently, the two wave functions may differ only in phase. Hence,

$$\Psi(a, b) = \exp(i\alpha)\Psi(b, a) \tag{5.60}$$

where $\exp(i\alpha)$ is a phase factor of modulus unity. The phase shift must have been produced by the operation which interchanged the two particles. If the operation is repeated, the phase change also will be repeated, so that

$$\Psi(a, b) = \exp(i\alpha)\Psi(b, a) = \exp(2i\alpha)\Psi(a, b) \tag{5.61}$$

and hence

$$\exp(i\alpha) = \pm 1. \tag{5.62}$$

From this analysis it is concluded that for a system of two identical particles either

$$\Psi(a, b) = \Psi(b, a) \tag{5.63}$$

in which case it is said that the wave function of the system is symmetrical with respect to the interchange of the coordinates of the two particles; or

$$\Psi(a, b) = -\Psi(b, a) \tag{5.64}$$

in which case it is said that the wave function of the system is antisymmetrical with respect to the interchange of the coordinates of the two particles. This argument is equally valid for assemblies of more than two particles, for the two particles to be exchanged may be labelled 1 and 2, and the remainder 3, 4, etc., at positions c, d, etc., remain unchanged and so appear in the same position in all the above equations.

In the most general sense, all the particles of a particular kind in the universe constitute a system; and the wave functions describing this grand assembly are also subject to the conditions (5.63) or (5.64). The total wave function will be made up of the functions for all the individual systems and since the superimposition of a symmetrical and an antisymmetrical function is asymmetrical, satisfying neither of the relations above, it follows that all the wave functions of systems of particles of a given type must have the same symmetry. Since an alteration in the symmetry would consequently involve a simultaneous change throughout the whole universe, it also follows that the symmetry does not change with time. Consequently, the type of symmetry is a property of that kind of particle. Experimentally, it is found that particles with half integer spin are described by antisymmetrical wave functions, while particles with interger spin are described by symmetrical wave functions. In the former class are electrons, protons, neutrons and atoms composed of odd numbers of elementary particles: in the latter are atoms and molecules composed of even numbers of elementary particles.

This property of symmetry produces an important consequence. Each particle in an assembly may be regarded as a single-particle system, described by a single-particle wave function. The total wave function for the system must then be some combination of the single-particle functions. Suppose in a two-particle system one particle with coordinates a is in a state labelled by a set of quantum numbers (including spin) p; and the second particle with coordinates b is in a state labelled by a set of quantum numbers q. The single particle wave functions corresponding to this arrangement are $\Psi_p(a)$ and $\Psi_q(b)$. However, it is also possible that the particle with coordinates a is in state q, and the particle with coordinates b in state p. The single-particle wave functions corresponding to this arrangement are $\Psi_p(b)$ and $\Psi_q(a)$. The two-particle wave function describing the system in which one particle is in a state p and another in a state q may be written

Ψ_{pq}, and this must be equal to some combination of the four single-particle wave functions above. Furthermore, the combination must be chosen so that the resulting two-particle wave function has the required symmetry. It is easy to write down a symmetrical combination:

$$\Psi_{pq} = \Psi_p(a)\Psi_q(b)+\Psi_p(b)\Psi_q(a) \tag{5.65}$$

it can be seen that if the position coordinates a and b are reversed, the wave function (5.64) is unchanged. An antisymmetrical combination can also be directly written:

$$\Psi_{pq} = \Psi_p(a)\Psi_q(b)-\Psi_p(b)\Psi_q(a) \tag{5.66}$$

and it can be seen that if a and b are reversed in (5.66) the function changes sign. The functions (5.65) and (5.66) can easily be generalised to a system of n particles with position coordinates $a, b, c, \ldots,$ in states $p, q, r, \ldots.$ For the symmetrical case:

$$\Psi_s = \sum_P \Psi_p(a)\Psi_q(b)\Psi_r(c) \cdots \tag{5.67}$$

and for the antisymmetrical:

$$\Psi_a = \sum \pm \Psi_p(a)\Psi_q(b)\Psi_r(c)\ldots \tag{5.68}$$

In (5.67) the summation is taken over all possible permutations of the $a, b, c, \ldots,$ among the states $p, q, r, \ldots;$ in (5.68) the summation is also taken over all permutations, but with the additional stipulation that if the permutation is produced from the original arrangement by an *odd* number of particle interchanges, the negative sign shall be taken: if by an *even* number of interchanges, the positive sign. The summation (5.68) is then the expansion of the determinant

$$\Psi_a = \begin{vmatrix} \Psi_p(a) & \Psi_p(b) & \Psi_p(c)\ldots \\ \Psi_q(a) & \Psi_q(b) & \Psi_q(c)\ldots \\ \Psi_r(a) & \Psi_r(b) & \ldots \\ \vdots & \vdots & \end{vmatrix} \tag{5.69}$$

It is well known that if two columns of a determinant are interchanged, the determinant changes sign, so that the determinant does have the required property of antisymmetry. However, determinants also have the property that if two rows are equal the determinant vanishes. Since this is contrary to the conditions of good behaviour, it follows that in fact no two rows of the determinant (5.69) can be equal, and hence no two particles can be represented by the same wave function at the same time, or, in other terms, no two particles can be in the same state, specified by a given set of four quantum numbers. This is a statement of the Pauli exclusion principle, already introduced on empirical grounds for electrons in Chapter 2. In fact, the principle applies to system of all particles of half-integer spin. It can be seen that it does not apply to systems of particles of integer spin, since the function (5.67) does not vanish if two of the states p, q, r ..., are set equal. It should be noted that none of the above wave functions have been normalised.

CHAPTER 6

A Wave-Mechanical Treatment of the Simple Harmonic Oscillator and the Hydrogen Atom

The Simple Harmonic Oscillator

In this chapter two rather more complicated problems are examined, partly because the results are of value in studying the solid state, and partly to demonstrate the sort of algebraic techniques that may be used for solving more complicated Schrödinger equations. First the possible energy levels of a simple harmonic oscillator will be determined. A three-dimensional simple harmonic oscillator can be separated into three identical one-dimensional oscillators, so that it is only necessary to consider the one-dimensional case.

Consider a particle mass m executing simple harmonic motion about a central point with frequency ν, the displacement from the central point being x. The potential energy is then proportional to the square of the displacement:

$$V = \tfrac{1}{2} f x^2. \tag{6.1}$$

The force acting on the particle is

$$F = -\frac{dV}{dx} = -fx. \tag{6.2}$$

The negative sign indicates that the force tends to restore the particle to the central position. The equation of motion of the particle is then

$$m\frac{d^2x}{dt^2} = -fx. \tag{6.3}$$

The general solution of this equation may be written

$$x = a\cos[(f/m)^{\frac{1}{2}}t+\varepsilon] \tag{6.4}$$

where a is an arbitrary constant determining the amplitude of the vibration and ε is an arbitrary constant determining its phase.

Equation (6.4) is unchanged by the addition of 2π to the factor inside the square brackets, so the motion repeats itself in a period of time τ, where

$$(f/m)^{\frac{1}{2}}\tau = 2\pi. \tag{6.5}$$

The frequency ν is then

$$\nu = \frac{1}{\tau} = \left(\frac{f}{4\pi^2 m}\right)^{\frac{1}{2}}. \tag{6.6}$$

The total energy of the particle is equal to the sum of the potential and kinetic energies:

$$E = \tfrac{1}{2}m\left(\frac{dx}{dt}\right)^2 + \tfrac{1}{2}fx^2. \tag{6.7}$$

Substituting from (6.4),

$$\begin{aligned} E &= \tfrac{1}{2}m[2\pi\nu a\sin(2\pi\nu t+\varepsilon)]^2 + \tfrac{1}{2}(2\pi\nu)^2 m[a\cos(2\pi\nu t+\varepsilon)]^2 \\ &= 2m(\pi\nu a)^2\{\cos^2(2\pi\nu t+\varepsilon)+\sin^2(2\pi\nu t+\varepsilon)\} \\ &= 2m(\pi\nu a)^2. \end{aligned} \tag{6.8}$$

The Hamiltonian of the system corresponding to (6.7) is

$$H = -\frac{\hbar^2}{2m}\frac{\partial^2}{\partial x^2} + \tfrac{1}{2}fx^2 \tag{6.9}$$

and the Scrödinger equation for the system is

$$\frac{\partial^2\psi}{\partial x^2} + \frac{2m}{\hbar^2}(E - \tfrac{1}{2}fx^2)\psi = 0. \tag{6.10}$$

An approximate solution of this equation may be written for the case where x is large, so that E is small in comparison to $\frac{1}{2}fx^2$. Equation (6.10) may then be written

$$\frac{\partial^2\psi_0}{\partial x^2} - \frac{mf}{\hbar^2}x^2\psi_0 = 0 \tag{6.11}$$

or

$$\left[\frac{\partial}{\partial x} - \sqrt{\left(\frac{mf}{\hbar^2}\right)}x\right]\left[\frac{\partial}{\partial x} + \sqrt{\left(\frac{mf}{\hbar^2}\right)}x\right]\psi_0 = 0 \tag{6.12}$$

so that

$$\frac{\partial\psi_0}{\partial x} = \pm\sqrt{\left(\frac{mf}{\hbar^2}\right)}x\psi_0$$

$$\psi_0 = C\exp\pm\left[\frac{1}{2}\sqrt{\left(\frac{mf}{\hbar^2}\right)}x^2\right] = C\exp\pm\left(\frac{\pi\nu m}{\hbar}x^2\right) \tag{6.13}$$

where C is a constant. The standard boundary condition eliminates the positive sign, so the solution

$$\psi_0 = C\exp-\left(\frac{\pi\nu m}{\hbar}x^2\right) \tag{6.14}$$

is the limiting form of the solution of (6.10) for large x. Differentiating (6.14) twice,

$$\frac{d^2\psi_0}{dx^2} = \left(\frac{2\pi\nu m}{\hbar}\right)^2 x^2\psi_0 - \frac{2\pi\nu m}{\hbar}\psi_0 \tag{6.15}$$

and

$$\frac{d^2\psi_0}{dx^2} - \left[\frac{mf}{\hbar^2}x^2 - \sqrt{\left(\frac{mf}{\hbar^2}\right)}\right]\psi_0 = 0 \tag{6.16}$$

which may be written

$$\frac{d^2\psi_0}{dx^2} + \frac{2m}{\hbar^2}\left[\frac{h\nu}{2} - \frac{fx^2}{2}\right]\psi_0 = 0. \tag{6.17}$$

Comparing (6.17) with (6.10), it follows that the energy associated with this limiting solution is

$$E_0 = \tfrac{1}{2}h\nu. \tag{6.18}$$

Using (6.8) and the result that $f = 2E/a^2$, then $f = \hbar^2/ma^4$ and so $2\pi^2 \nu m/h = \tfrac{1}{2}a^2$. The asymptotic solution (6.14) may therefore be written

$$\psi_0 = C \exp\left[-\frac{1}{2}\left(\frac{x}{a}\right)^2\right]. \tag{6.19}$$

The solution for smaller x may be approximated by replacing the constant C in (6.19) with a slowly varying function of x/a. Writing $s = x/a$,

$$\psi = u(s) \exp(-\tfrac{1}{2}s^2). \tag{6.20}$$

Differentiating this trial solution,

$$\frac{d\psi}{ds} = \exp(-\tfrac{1}{2}s^2)\left[-us + \frac{du}{ds}\right]$$

and

$$\frac{d^2\psi}{ds^2} = \exp(-\tfrac{1}{2}s^2)\left[(s^2-1)u - 2s\frac{du}{ds} + \frac{d^2u}{ds^2}\right].$$

Now $\exp(-\tfrac{1}{2}s^2) = \psi/u$, so

$$\frac{d^2\psi}{ds^2} = \frac{\psi}{u}\left(\frac{d^2u}{ds^2} - 2s\frac{du}{ds}\right) + (s^2-1)\psi \tag{6.21}$$

but

$$\frac{d^2\psi}{dx^2} = \frac{1}{a^2}\frac{d^2\psi}{ds^2},$$

so substituting in (6.10) and multiplying through by d^2u,

$$\frac{d^2u}{ds^2} - 2s\frac{du}{ds} + \left[\frac{2ma^2}{\hbar^2}(E - \tfrac{1}{2}fa^2s^2) + s^2 - 1\right]u = 0.$$

The coefficient of s^2 in the bracket is

$$\left(1-\frac{2ma^2}{\hbar^2}\cdot\tfrac{1}{2}fa^2\right)=0,$$

so the equation becomes

$$\frac{d^2u}{ds^2}-2s\frac{du}{ds}+\left(\frac{2E}{h\nu}-1\right)u=0. \tag{6.22}$$

Now, since u is a slowly varying function of s, it may be replaced by a polynomial in s:

$$u=a_0+a_1s+a_2s^2+\ldots+a_{n-1}s^{n-1}+a_ns^n. \tag{6.23}$$

Differentiating and rearranging:

$$\frac{d^2u}{ds^2}=2a_2+6a_3s+\cdots+(n-1)(n-2)a_{n-1}s^{n-3}+n(n-1)a_ns^{n-2};$$

$$-2s\frac{du}{ds}=-2a_1s-\cdots-2(n-1)a_{n-1}s^{n-1}-2na_ns^n;$$

$$\left(\frac{2E}{h\nu}-1\right)u=\left(\frac{2E}{h\nu}-1\right)a_0+\cdots+\left(\frac{2E}{h\nu}-1\right)a_{n-1}s^{n-1}+\left(\frac{2E}{h\nu}-1\right)a_ns^n.$$

Summing the coefficients of each power of s separately,

$$0=\left[2a_2+\left(\frac{2E}{h\nu}-1\right)a_0\right]+\left[6a_3-2a_1+\left(\frac{2E}{h\nu}-1\right)a_1\right]s+\cdots$$
$$+\left[\left(\frac{2E}{h\nu}-1\right)a_{n-1}-2(n-1)a_{n-1}\right]s^{n-1}$$
$$+\left[\left(\frac{2E}{h\nu}-1\right)a_n-2na_n\right]s^n. \tag{6.24}$$

Since (6.24) holds for all s, each coefficient must separately equal zero. The coefficient of s^n is

$$a_n\left[\frac{2E}{h\nu}-1-2n\right]$$

and neglecting the trivial result $a_n = 0$,

$$E = (n+\tfrac{1}{2})h\nu = (n+\tfrac{1}{2})\hbar\omega. \tag{6.25}$$

The polynomial (6.23) terminates; if this were not so u, and hence ψ, would become infinite, and so not be well behaved. The energy of a simple harmonic oscillator is thus quantised, and it is interesting to note that zero energy is not an eigenvalue of the Hamiltonian: the lowest energy, the so-called zero-point energy, is $\frac{1}{2}\hbar\omega$.

A Particle in a Coulomb Field: the Hydrogen Atom

The hydrogen atom consists of an electron, a light particle with a negative charge, and a proton, a relatively massive particle with an equal positive charge. As it happens, the sizes of the two particles (so far as one can talk about the "size" of a fundamental particle) are very small in comparison to the size of the atom. It is not therefore unreasonable to discuss the hydrogen atom as though it were a point negative charge of mass m moving in the Coulomb field due to a point positive charge of infinite mass. The energy of interaction is dependent only on the distance between the two charges, and not on the direction of one particle with respect to the other. The potential field is therefore spherically symmetrical, and it is appropriate to write the Laplacian in terms of spherical coordinates, as in (4.31). If the two particles have charges of equal magnitude e but of opposite sign, and are separated by a distance r, the potential V becomes

$$V(r,\theta,\phi) = -\frac{e^2}{r}. \tag{6.26}$$

The Schrödinger equation for the system is then

$$\left\{\frac{\partial^2}{\partial r^2}+\frac{2}{r}\frac{\partial}{\partial r}+\frac{1}{r^2\sin\theta}\frac{\partial}{\partial\theta}\left(\sin\theta\frac{\partial}{\partial\theta}\right)+\frac{1}{r^2\sin^2\theta}\frac{\partial^2}{\partial\phi^2}\right\}\psi + \frac{2m}{\hbar^2}\left(E+\frac{e^2}{r}\right)\psi = 0. \quad (6.27)$$

If (6.27) is multiplied by r^2 it may be separated into two parts, one of which is a function of r only, and the other of θ and ϕ. Accordingly, the wave function must also be separable:

$$\psi(r,\theta,\phi) = R(r)S(\theta,\phi). \quad (6.28)$$

Substituting the wave function (6.28) in (6.27) and multiplying by $r^2/\psi = r^2/RS$ we obtain:

$$\frac{r^2}{R}\frac{\partial^2 R}{\partial r^2}+\frac{2r}{R}\frac{\partial R}{\partial r}+\frac{1}{S\sin\theta}\frac{\partial}{\partial\theta}\left(\sin\theta\frac{\partial S}{\partial\theta}\right)+\frac{1}{S\sin^2\theta}\frac{\partial^2 S}{\partial\phi^2} + \frac{2r^2m}{\hbar^2}\left(E+\frac{e^2}{r}\right) = 0. \quad (6.29)$$

Since this equation has the form

$$f_1(r)+f_2(\theta,\phi) = 0 \quad (6.30)$$

it follows that

$$f_1(r) = K \quad \text{and} \quad f_2(\theta,\phi) = -K \quad (6.31)$$

where K is independent of r, θ and ϕ.
Consider f_1 first.

$$\frac{r^2}{R}\left\{\frac{\partial^2}{\partial r^2}+\frac{2}{r}\frac{\partial}{\partial r}+\frac{2m}{\hbar^2}\left(E+\frac{e^2}{r}\right)\right\} R = K. \quad (6.32)$$

Dividing by r^2/R,

$$\frac{\partial^2 R}{\partial r^2}+\frac{2}{r}\frac{\partial R}{\partial r}+\left(\frac{2m}{\hbar^2}E+\frac{2m}{\hbar^2}\cdot\frac{e^2}{r}-\frac{K}{r^2}\right)R = 0. \quad (6.33)$$

At small r, the coefficient of R approaches $-K/r^2$, so the equation becomes

$$r^2 \frac{\partial^2 R}{\partial r^2} + 2r \frac{\partial R}{\partial r} - KR = 0. \tag{6.34}$$

Setting $r = \exp z$,

$$\left(\frac{\partial^2}{\partial z^2} - \frac{\partial}{\partial z}\right) R + 2\frac{\partial}{\partial z} R - KR = 0$$

or

$$\frac{\partial^2}{\partial z^2} R + \frac{\partial}{\partial z} R - KR = 0. \tag{6.35}$$

The roots of the auxiliary equation are

$$-\tfrac{1}{2} \pm \sqrt{(\tfrac{1}{4} + K)}$$

and the two independent solutions are

$$z = l \quad \text{and} \quad z = -(l+1) \tag{6.36}$$

where

$$l = -\tfrac{1}{2} + \sqrt{(\tfrac{1}{4} + K)}.$$

The linearly independent solutions of (6.34) are thus

$$R = r^l \quad \text{and} \quad R = r^{-(l+1)} \tag{6.37}$$

and

$$K = l(l+1) \tag{6.38}$$

Equation (6.33) becomes

$$\frac{\partial^2}{\partial r^2} R + \frac{2}{r}\frac{\partial}{\partial r} R + \left\{-\alpha^2 + \frac{2\beta}{r} - \frac{l(l+1)}{r^2}\right\} R = 0 \tag{6.39}$$

where $2mE/\hbar^2 = -\alpha^2$, the negative sign indicating bound states only; and $2\beta = 2me^2/\hbar^2$.

As r becomes large (6.39) becomes

$$\frac{\partial^2}{\partial r^2} R = \alpha^2 R \tag{6.40}$$

the solution of which is

$$R = c \exp \pm(\alpha r). \tag{6.41}$$

The positive sign is eliminated by the standard boundary condition. As in the case of the simple harmonic oscillator, the solution for finite r is approached by replacing the constant c by a slowly varying function of r. The trial solution is then

$$R = u(r) \exp -(\alpha r). \tag{6.42}$$

Substituting in (6.39) and multiplying by u:

$$\frac{\partial^2}{\partial r^2} u + 2\left(\frac{1}{r} - \alpha\right)\frac{\partial}{\partial r} u + \left\{2\frac{\beta-\alpha}{r} - \frac{l(l+1)}{r^2}\right\} u = 0. \tag{6.43}$$

Since u is a slowly varying function of r, it may be replaced by a power series:

$$u = r^{\mu} \sum_{\nu=0}^{\infty} a_{\nu} r^{\nu} = r^{\mu}(a_0 + a_1 r + a_2 r^2 + \cdots + a\nu r^{\nu} + \cdots) \tag{6.44}$$

where ν is an integer, a_ν is a constant and μ may be a fraction. Differentiating and substituting in (6.43) a recurrence formula is obtained:

$$a_\nu\{(\mu+\nu)(\mu+\nu+1) - l(l+1)\} = 2a_{\nu-1}\{\alpha(\mu+\nu) - \beta\}. \tag{6.45}$$

The first coefficient in the series is a_0, so a_{-1} must be zero. Setting $\nu = 0$, the right-hand side of (6.45) vanishes, and

$$a_0\{\mu(\mu+1) - l(l+1)\} = 0 \tag{6.46}$$

so either

$$\mu = l \quad \text{or} \quad \mu = -(l+1). \tag{6.47}$$

When $\mu = -(l+1)$, then the coefficient a_ν becomes infinite when $\nu = 2l+1$. a_ν is always finite for $\mu = l$, so the second solution (6.47) is rejected. The recurrence formula (6.45) becomes

$$a_\nu\{(l+\nu)(l+\nu+1) - l(l+1)\} = 2a_{\nu-1}\{\alpha(l+\nu) - \beta\}. \tag{6.48}$$

The power series (6.44) becomes

$$u = a_0 r^l + a_0 \left\{\frac{\alpha(l+1)-\beta}{l+1}\right\} r^{l+1} + a_0 \left\{\left[\frac{\alpha(l+1)-\beta}{l+1}\right]\left[\frac{\alpha(l+2)-\beta}{l+3}\right]\right\} r^{l+2} + \cdots \tag{6.49}$$

At large α the recurrence formula (6.45) is approximately

$$\frac{a_\nu}{a_{\nu-1}} \cong \frac{2\alpha}{l+\nu+1} \tag{6.50}$$

but this is the same as the recurrence formula for the exponential series $\exp(+2\alpha r)$, so that at large α, $u(r) \simeq \exp(+2\alpha r)$ and consequently $R = u \exp(-\alpha r) \simeq \exp(+\alpha r)$. But this function does not satisfy the Standard Boundary Condition, and consequently the series (6.49) must terminate. Suppose there are ν terms in the polynomial. Then $a_\nu = 0$, but $a_{\nu-1} \neq 0$.
From (6.48) it follows that

$$\alpha(l+\nu) = \beta.$$

Writing $l+\nu = n$, then $\alpha n = \beta$, or

$$E = -\frac{me^4}{2\hbar^2} \cdot \frac{1}{n^2}. \tag{6.51}$$

Equation (6.51) is identical with (2.30) of the Bohr theory, provided that n is an integer. ν is the number of terms in the series, and hence is a non-zero positive integer, but it remains to be demonstrated that l is integral.

Consider the function $f_2(\theta, \phi)$ of (6.30). This may be written

$$\frac{1}{\sin\theta}\frac{\partial}{\partial\theta}\left(\sin\theta\frac{\partial}{\partial\theta}s\right) + \frac{1}{\sin^2\theta}\frac{\partial^2}{\partial\phi^2}s + l(l+1)s = 0. \tag{6.52}$$

This equation may be further separated. Writing

$$s(\theta, \phi) = \Theta(\theta)\Phi(\phi) \tag{6.53}$$

and substituting in (6.52) and multiplying by $\sin^2\theta/\Theta\Phi$, the two equations

$$\frac{\sin\theta}{\Theta}\frac{\partial}{\partial\theta}\left(\sin\theta\frac{\partial}{\partial\theta}\Theta\right) + l(l+1)\sin^2\theta = K \tag{6.54}$$

and

$$\frac{1}{\Phi}\frac{\partial^2}{\partial\phi^2}\Phi = -K \tag{6.55}$$

are obtained, where again K is a constant independent of r, θ and ϕ. Equation (6.55) may be integrated directly:

$$\Phi = A \exp(\pm i m_l \phi) \tag{6.56}$$

where A is a constant and $m_l^2 = K$. Since the motion is periodic,

$$\Phi(\phi) = \Phi(\phi + 2\pi) \tag{6.57}$$

and hence m_l is an integer or zero. The normalising condition is

$$\int_0^{2\pi} \Phi\Phi^* d\phi = 1$$

giving

$$A = 1/2\pi$$

so the solution of (6.55) may finally be written

$$\Phi = \frac{1}{\sqrt{(2\pi)}} \exp(i m_l \phi) \tag{6.58}$$

where m_l is any positive or negative integer or zero. The component of momentum p_ϕ is the angular component of momentum about one Cartesian coordinate. Using the operator associated in the Schrödinger scheme with momentum, $-i\hbar(\partial/\partial\phi)$,

$$p_\phi \psi = -i\hbar \frac{\partial}{\partial\phi} \psi. \tag{6.59}$$

Now $\psi = \text{const.}\ R(r)\Theta(\theta) \exp(i m_l \phi)$, so

$$-i\hbar \frac{\partial}{\partial\phi} \psi = (-i\hbar)(i m_l \psi). \tag{6.60}$$

Hence

$$p_\phi = m_l \hbar. \tag{6.61}$$

This is identical with the Bohr quantum condition (2.46). Equation (6.54) becomes

$$\frac{1}{\Theta} \frac{1}{\sin\theta} \frac{\partial}{\partial\theta}\left(\sin\theta \frac{\partial}{\partial\theta} \Theta\right) - \frac{m_l^2}{\sin^2\theta} = -l(l+1). \tag{6.62}$$

The solution of this equation is

$$\Theta = \sin^m \theta \, P_l^m(\cos \theta) \tag{6.63}$$

where $P_l^m(\cos \theta)$ is the Legendre associated function

$$P_l^m(\cos \theta) = \frac{1}{2^l l!} \frac{\partial^{m+l}}{\partial \mu^{m+l}} (\mu^2 - 1)^l \tag{6.64}$$

where $\mu = \cos \theta$ and $m = | m_l |$.

(*Higher Mathematics for Engineers and Physicists*, I. S. and E. S. Sokolnikoff, McGraw-Hill; and *Special Functions of Physics and Chemistry*, I. N. Sneddon, Oliver & Boyd.)

The highest term in the bracket in (6.64) is μ^{2l}. This term has to be differentiated $m+l$ times, and consequently vanishes if $m = | m_l |$ is greater than l. The range of values of m_l is thus $-l \gtreqless m_l \gtreqless l$. Finally, the standard boundary condition requires that (6.64) shall converge, and this is only possible if the series terminates. Consequently, l must be a positive integer or zero. This proves that the quantity n introduced in (6.51) is identical with the quantum number n of the Bohr theory. It has been demonstrated that the integer m_l appearing in (6.56) is identical with the magnetic quantum number of the Bohr theory, and the limitation on the values of m_l introduced above suggests that the integer l is related to the angular quantum number k of the old quantum theory. The permitted range of values of l is different from that of k, however, so it is interesting to determine the relation of the quantum number l to the angular component of momentum in the wave-mechanical model.

The angular momentum p_θ can be calculated for the special case of circular orbits quite easily. Setting $r = a$, where a is a constant, in (6.27) produces the Schrödinger equation

$$\frac{1}{a^2}\frac{\partial^2}{\partial \theta^2}\psi + \frac{\cot \theta}{a^2}\frac{\partial}{\partial \theta}\psi + \frac{1}{a^2 \sin \theta}\frac{\partial^2}{\partial \phi^2}\psi + \frac{2m}{\hbar^2}(E - V_0)\psi = 0 \tag{6.65}$$

where the potential energy $V_0 = e^2/a$. This is identical with (6.52) provided that

$$\frac{2ma^2}{\hbar^2}(E-V_0) = l(l+1). \tag{6.66}$$

But $E-V_0$ represents the purely kinetic angular component of energy

$$= \frac{\hbar^2}{2ma^2}\,l(l+1)$$

$$= p_\theta/2ma^2.$$

Therefore,

$$p_\theta = \hbar\sqrt{[l(l+1)]}. \tag{6.67}$$

This differs slightly from the Bohr quantum condition (2.22), but it is clear that the integer l occupies the same position in the new quantum theory that the quantum number k occupied in the Bohr theory.

The Labelling of States and the Periodic Table

In the absence of a magnetic field, the energy of a state is determined solely by the principal quantum number n and the angular quantum number l. These two quantum numbers are used to label "shells", composed of a number of states having the same energy. For a given value of l there are $2l+1$ possible values of m_l, and for each of these there are two possible values of the spin quantum number m_s. The total number of states in each shell is thus $2(2l+1)$. For convenience in labelling, the principal quantum number of each shell is indicated by a number, and the angular quantum number by a letter. The letters are derived from spectroscopic terminology:

$l = 0 = s$ from the spectroscopic term "sharp",

$l = 1 = p$ from the spectroscopic term "principal",

$l = 2 = d$ from the spectroscopic term "diffuse",

$l = 3 = f$ from the spectroscopic term "faint".

The shells are thus

$1s$ comprising two states,
$2s$ comprising two states,
$2p$ comprising six states,
$3s$ comprising two states,
$3p$ comprising six states,
$3d$ comprising ten states,
$4s$ comprising two states,
etc.

The energy of the states in many-electron atoms does not in fact increase in this simple way. The contribution of the second quantum number to the energy increases as the total energy increases, so that in fact the $4s$ shell has lower energy than the $3d$ shell. The major energy discontinuities are between the $1s$ and $2s$ shells, between the $2p$ and $3s$ shells, and between the $3p$ and $4s$ shells. Because of the Pauli exclusion principle, as the number of electrons in an atom increases, the shells are gradually filled up. Hydrogen thus has one $1s$ electron; helium has two $1s$ electrons, and consequently has a full $1s$ shell. Lithium has a full $1s$ shell, and one $2s$ electron. Because of the difference in energy between the $1s$ and $2s$ states, the helium atom has a special stability: it is unwilling either to accept or to lose an electron; whereas the lithium atom easily gives up its single $2s$ electron forming a positive "ion" which has two electrons in the $1s$ shell, the same electron structure as helium. In practice, these properties are exhibited by the chemical inertness of helium, and the extreme reactivity of lithium, which exhibits a valency of one, and is strongly electropositive. The next element is beryllium, which has two $1s$ electrons and two $2s$ electrons. Because there is not a great difference of energy between the $2s$ and $2p$ shells this structure does not have any special stability. The beryllium atom will give up both its electrons to form a doubly charged or "divalent" positive ion, but because of the greater charge on the nucleus this is harder to do than with lithium, so that the reactivity of beryllium is considerably less than that of lithium. Boron has a full 1*s* shell, a full 2*s* shell, and one 2*p* electron. It exhibits a valency of three under some circumstances, but it cannot be described as a very reactive material. Carbon is the next element, with two 2*p* electrons. It exhibits a valency of four, and can both donate electrons to form a positively charged "cation" or accept electrons to form a negatively charged "anion". Nitrogen and oxygen have three and four 2*p* electrons, respectively; fluorine has five, and neon has six, thus filling the 2*p* shell. There is a large energy discontinuity between the 2*p* and 3*s* shells, and like helium, the electronic structure of neon has very high stability. Neon itself is the second member of the inert gases. Fluorine readily accepts an electron to achieve the neon electronic structure as a monovalent anion, and is consequently very reactive. Oxygen accepts two electrons to form a divalent anion, but again the excess charge makes this ion rather less stable, so that the reactivity is less. Nitrogen accepts three electrons, but again the reactivity is decreased because of the reduction in stability due to the additional charge on the ion. The next element after neon is sodium, which has the electronic structure of neon plus a single electron in the 3*s* shell. Exactly as for lithium, sodium is extremely reactive, readily giving up the extra electron to form a monovalent cation with the electron structure of neon. Chemically, therefore, sodium is very similar to lithium, and these two elements are the first members of group IA of the Periodic Table, the alkali metals. The other members are potassium, caesium and rubidium, and each has one electron more than an inert gas. The next element is magnesium, with two 3*s* electrons: chemically, this is similar to beryllium, forming a divalent cation. Beryllium and magnesium are the first two members of group IIB of the Periodic Table. These are divalent, and are generally rather less reactive than the alkali metals. Other members of the group are zinc, cadmium and mercury. The following elements fill up the 3*p* shell: chlorine with five 3*p* electrons has properties similar to those of fluorine, and is the second member of group VIIB, the halogens. Argon, with six 3*p* electrons is an inert gas. The next element has the argon electrons plus one in the 4*s* shell: it is potassium and is an alkali metal. The next element is calcium with two 4*s* electrons: it is chemically similar to magnesium and is the first element of group IIA, the alkaline earths. Sometimes beryllium and magnesium are included in this group. The next element is scandium: in addition to the full shells of argon it has two 4*s* electrons and one 3*d* electron: this arrangement is not parallel to any of the preceding elements. The next few elements gradually fill the 3*d* shell: titanium has two 3*d* electrons, vanadium three; chromium, however, has five 3*d* electrons and only one 4*s* electron. This arrangement arises because as each shell fills up the electrons attempt to occupy states so that the number of unpaired spins is a maximum (Hund's rule): for example, there are three 3*p* states with two electrons of opposite spin in each. Phosphorus, which has three 3*p* electrons, has one in each of the three states, rather than two in one state, one in another, with the third empty. The "half-full shell" with one electron in each doubly degenerate state has a stability similar to, but very much less marked than, the stability of a full shell. This special stability is sufficient to cause an electron to be "pulled down" from the 4*s* shell in chromium. Manganese also has five 3*d* electrons, but two 4*s* electrons. Iron has six 3*d* electrons, cobalt seven, and nickel eight. Because of the stability of a full shell, the next element, copper, has ten 3*d* electrons and one 4*s* electron, and the metal is predominantly a monovalent metal, like the alkali metals, although the reactivity is much less. Copper is consequently regarded as the first member of a group labelled IB, the noble metals. The other two members of this group are silver and gold. The eight elements from scandium to nickel which gradually fill the 3*d* shell are called transition elements, and this period is called the first long period. In Table 6.1 the electron configurations of the elements are shown, and in Table 6.2 the periodic classification of the elements is shown.

In terms of the wave theory, it is no longer permissible to

TABLE 6.1. THE ELECTRONIC CONFIGURATION OF NEUTRAL ATOMS OF THE ELEMENTS. THE ASSIGNMENTS FOR THE 4*f* AND 5*f* SERIES OF RARE EARTHS ARE UNCERTAIN

Element				Quantum numbers									
Number	Name	Symbol	Weight	1	2		3			4			
				s	*s*	*p*	*s*	*p*	*d*	*s*	*p*	*d*	*f*
1	Hydrogen	H	1·008	1									
2	Helium	He	4·003	2									
3	Lithium	Li	6·939	2	1								
4	Beryllium	Be	9·012	2	2								
5	Boron	B	10·811	2	2	1							
6	Carbon	C	12·011	2	2	2							
7	Nitrogen	N	14·007	2	2	3							
8	Oxygen	O	15·999	2	2	4							
9	Fluorine	F	18·998	2	2	5							
10	Neon	Ne	20·183	2	2	6							
11	Sodium	Na	22·990	2	2	6	1						
12	Magnesium	Mg	24·312	2	2	6	2						
13	Aluminium	Al	26·981	2	2	6	2	1					
14	Silicon	Si	28·086	2	2	6	2	2					
15	Phosphorous	P	30·974	2	2	6	2	3					
16	Sulphur	S	32·064	2	2	6	2	4					
17	Chlorine	Cl	35·453	2	2	6	2	5					
18	Argon	Ar	39·948	2	2	6	2	6					
19	Potassium	K	39·102	2	2	6	2	6		1			
20	Calcium	Ca	40·08	2	2	6	2	6		2			
21	Scandium	Sc	44·956	2	2	6	2	6	1	2			
22	Titanium	Ti	47·90	2	2	6	2	6	2	2			
23	Vanadium	V	50·942	2	2	6	2	6	3	2			
24	Chromium	Cr	51·996	2	2	6	2	6	5	1			
25	Manganese	Mn	54·938	2	2	6	2	6	5	2			
26	Iron	Fe	55·847	2	2	6	2	6	6	2			
27	Cobalt	Co	58·933	2	2	6	2	6	7	2			
28	Nickel	Ni	58·71	2	2	6	2	6	8	2			
29	Copper	Cu	63·54	2	2	6	2	6	10	1			
30	Zinc	Zn	65·37	2	2	6	2	6	10	2			
31	Gallium	Ga	69·72	2	2	6	2	6	10	2	1		
32	Germanium	Ge	72·59	2	2	6	2	6	10	2	2		
33	Arsenic	As	74·922	2	2	6	2	6	10	2	3		
34	Selenium	Se	78·96	2	2	6	2	6	10	2	4		
35	Bromine	Br	79·909	2	2	6	2	6	10	2	5		
36	Krypton	Kr	83·80	2	2	6	2	6	10	2	6		

TABLE 6.1 (cont.)

Element				Quantum numbers														
Number	Name	Symbol	Weight	1	2	3	4				5					6		
							s	*p*	*d*	*f*	*s*	*p*	*d*	*f*	*g*	*s*	*p*	*d*
37	Rubidium	Rb	85·47	2	8	18	2	6			1							
38	Strontium	Sr	87·62	2	8	18	2	6			2							
39	Yttrium	Y	88·905	2	8	18	2	6	1		2							
40	Zirconium	Zr	91·22	2	8	18	2	6	2		2							
41	Niobium	Nb	92·906	2	8	18	2	6	4		1							
42	Molybdenum	Mo	95·94	2	8	18	2	6	5		1							
43	Technetium	Tc	[99]	2	8	18	2	6	6		1							
44	Ruthenium	Ru	101·07	2	8	18	2	6	7		1							
45	Rhodium	Rh	102·905	2	8	18	2	6	8		1							
46	Palladium	Pd	106·4	2	8	18	2	6	10									
47	Silver	Ag	107·870	2	8	18	2	6	10		1							
48	Cadmium	Cd	112·40	2	8	18	2	6	10		2							
49	Indium	In	114·82	2	8	18	2	6	10		2	1						
50	Tin	Sn	118·69	2	8	18	2	6	10		2	2						
51	Antimony	Sb	121·75	2	8	18	2	6	10		2	3						
52	Tellurium	Te	127·60	2	8	18	2	6	10		2	4						
53	Iodine	I	126·904	2	8	18	2	6	10		2	5						
54	Xenon	Xe	131·30	2	8	18	2	6	10		2	6						
55	Cesium	Cs	132·905	2	8	18	2	6	10		2	6				1		
56	Barium	Ba	137·34	2	8	18	2	6	10		2	6				2		
57	Lanthanum	La	138·91	2	8	18	2	6	10		2	6	1			2		
58	Cerium	Ce	140·12	2	8	18	2	6	10	2	2	6				2		
59	Praseodymium	Pr	140·907	2	8	18	2	6	10	3	2	6				2		
60	Neodymium	Nd	144·24	2	8	18	2	6	10	4	2	6				2		
61	Promethium	Pm	[145]	2	8	18	2	6	10	5	2	6				2		
62	Samarium	Sm	150·35	2	8	18	2	6	10	6	2	6				2		
63	Europium	Eu	151·96	2	8	18	2	6	10	7	2	6				2		
64	Gadolinium	Gd	157·25	2	8	18	2	6	10	7	2	6	1			2		
65	Terbium	Tb	158·924	2	8	18	2	6	10	8	2	6	1			2		
66	Dysprosium	Dy	162·50	2	8	18	2	6	10	10	2	6				2		
67	Holmium	Ho	164·930	2	8	18	2	6	10	11	2	6				2		
68	Erbium	Er	167·26	2	8	18	2	6	10	12	2	6				2		
69	Thulium	Tm	168·934	2	8	18	2	6	10	13	2	6				2		
70	Ytterbium	Yb	173·04	2	8	18	2	6	10	14	2	6				2		
71	Lutetium	Lu	174·97	2	8	18	2	6	10	14	2	6	1			2		
72	Hafnium	Hf	178·49	2	8	18	2	6	10	14	2	6	2			2		
73	Tantalum	Ta	180·948	2	8	18	2	6	10	14	2	6	3			2		

TABLE 6.1 (cont.)

Element				Quantum numbers															
Number	Name	Symbol	Weight	1	2	3	4	5					6				7	
								s	*p*	*d*	*f*	*g*	*s*	*p*	*d*	*f,g,h*	*s*	*p*
74	Tungsten	W	183·85	2	8	18	32	2	6	4			2					
75	Rhenium	Re	186·2	2	8	18	32	2	6	5			2					
76	Osmium	Os	190·2	2	8	18	32	2	6	6			2					
77	Iridium	Ir	192·2	2	8	18	32	2	6	9								
78	Platinum	Pt	195·09	2	8	18	32	2	6	9			1					
79	Gold	Au	196·967	2	8	18	32	2	6	10			1					
80	Mercury	Hg	200·59	2	8	18	32	2	6	10			2					
81	Tellurium	Tl	204·37	2	8	18	32	2	6	10			2	1				
82	Lead	Pb	207·19	2	8	18	32	2	6	10			2	2				
83	Bismuth	Bi	208·980	2	8	18	32	2	6	10			2	3				
84	Polonium	Po	[210]	2	8	18	32	2	6	10			2	4				
85	Astatine	At	[210]	2	8	18	32	2	6	10			2	5				
86	Radon	Rn	[222]	2	8	18	32	2	6	10			2	6				
87	Francium	Fr	[223]	2	8	18	32	2	6	10			2	6			1	
88	Radium	Ra	[226]	2	8	18	32	2	6	10			2	6			2	
89	Actinium	Ac	[227]	2	8	18	32	2	6	10			2	6	1			
90	Thorium	Th	232·038	2	8	18	32	2	6	10			2	6	2		2	
91	Protactinium	Pa	[231]	2	8	18	32	2	6	10	2		2	6	1		2	
92	Uranium	U	238·03	2	8	18	32	2	6	10	3		2	6	1		2	
93	Neptunium	Np	[237]	2	8	18	32	2	6	10	5		2	6			2	
94	Plutonium	Pu	[242]	2	8	18	32	2	6	10	6		2	6			2	
95	Americium	Am	[243]	2	8	18	32	2	6	10	7		2	6			2	
96	Curium	Cm	[247]	2	8	18	32	2	6	10	7		2	6	1		2	
97	Berkelium	Bk	[249]	2	8	18	32	2	6	10	8		2	6	1		2	
98	Californium	Cf	[251]	2	8	18	32	2	6	10	9		2	6	1		2	
99	Einsteinium	Es	[254]	2	8	18	32	2	6	10	10		2	6	1		2	
100	Fermium	Fm	[253]	2	8	18	32	2	6	10	11		2	6	1		2	
101	Mendelevium	Md	[256]	2	8	18	32	2	6	10	12		2	6	1		2	
102	Nobelium	No	[254]	2	8	18	32	2	6	10	13		2	6	1		2	
103	Lawrencium	Lw	[257]	2	8	18	32	2	6	10	14		2	6	1		2	
104				2	8	18	32	2	6	10	14		2	6	2		2	

TABLE 6.2. THE PERIODIC TABLE

I		II		III	IV	V	VI	VII	III	IV	V	VI	VII	VIII	IX	X	VIII	Group Number
A	B	A	B	B	B	B	B	B	A	A	A	A	A	A			B	
H 1																	He 2	Groups IIIA to X
Li 3			Be 4	B 5	C 6	N 7	O 8	F 9									Ne 10	are the "transition
Na 11			Mg 12	Al 13	Si 14	P 15	S 16	Cl 17									Ar 18	groups"
K 19		Ca 20							Sc 21	Ti 22	V 23	Cr 24	Mn 25	Fe 26	Co 27	Ni 28		
	Cu 29		Zn 30	Ga 31	Ge 32	As 33	Se 34	Br 35									Kr 36	
Rb 37		Sr 38							Y 39	Zr 40	Nb 41	Mo 42	Tc 43	Ru 44	Rh 45	Pd 46		
	Ag 47		Cd 48	In 49	Sn 50	Sb 51	Te 52	I 53									Xe 54	
Cs 55		Ba 56							La 57	Hf 72	Ta 73	W 74	Re 75	Os 76	Ir 77	Pt 78		
	Au 79		Hg 80	Tl 81	Pb 82	Bi 83	Po 84	At 85									Rn 86	
Fr 87		Ra 88							Ac 89	— 104								

Rare earths: type 4f	Ce 58	Pr 59	Nd 60	Pm 61	Sm 62	Eu 63	Gd 64	Tb 65	Dy 66	Ho 67	Er 68	Tm 69	Yb 70	Lu 71
type 5f	Th 90	Pa 91	U 92	Np 93	Pu 94	Am 95	Cm 96	Bk 97	Cf 98	Es 99	Fm 100	Md 101	No 102	Lw 103

discuss the shapes of orbits, since the concept of a localised electron performing specific orbits around the nucleus is specifically excluded in the formulation of quantum mechanics. However, it is possible to determine the probability that the electron is at a specific point in space, and this can be equated to the charge density at that point: in other words, the electron is in motion, since it possesses a momentum, but the details of its motion are inaccessible. A static distribution of probability can be determined, and over an interval of time this will be equal to the charge due to the fractional presence of the electron at each point in space. This "charge cloud" is a useful concept, and so far as an atom can be said to have a shape, it is the shape of the charge cloud of its electrons. The charge clouds having the greatest radial extent are those of the outermost electrons, which also determine the chemical valence of the atom. These electrons are called the valence electrons, and the remaining electrons, which are in full shells, and the nucleus are called the "ion core". Determination of the shape of the charge clouds due to electrons in various states is a problem in quantum mechanics which is rather beyond the scope of this book: the results alone will be quoted.

All full shells have spherically symmetrical charge clouds. It is therefore reasonable to think of inert gas atoms as spheres, and ions which have achieved the electron structure of inert gases by accepting or donating electrons may also be thought of as spheres. Half-full shells, with one electron in each doubly spin degenerate state, also have spherical symmetry. All *s* states consequently have spherical symmetry, since an *s* state can accommodate only two electrons, and so must be either half full or full. The three *p* states are mutually perpendicular.

The preceding discussion of electron structure applies only to free atoms: in chemical combinations the electron structure is obviously modified, but in fact any association, as molecules for example, or in a solid or a liquid, may modify the distribution of electrons among the states. The relative energies of the states may be modified, and even the structure of the states may be altered by the interaction between atoms. In the next chapter some of these modifications will be examined.

Spin and Wave Mechanics

The wave mechanics which has been outlined in the last two chapters does not predict a quantum number analogous to the spin quantum number m_s of the empirical theory. This is because the Schrödinger equation as set out here is capable only of dealing with translational degrees of freedom, and it is difficult to envisage a formulation of the theory which would explicitly take account of internal degrees of freedom. However, when the Schrödinger equation is written in accordance with the requirements of the theory of relativity, a fourth quantum number having the properties of m_s is produced. In this book we will use a non-relativistic quantum mechanics throughout, remembering the spin degeneracy of the states predicted in this way.

CHAPTER 7

Assemblies of Atoms

The Hydrogen Molecular Ion H_2^+

The simplest example of an assembly of atoms is the hydrogen molecular ion H_2^+, consisting of two protons and one electron. If the relative motion of the two protons is neglected, the problem may be regarded as the behaviour of the electron in two overlapping coulomb fields. At very large separations, the two coulomb fields may be considered to be independent, and the electron will be certainly associated with one proton or the other, since the probability of the electron transferring from one proton to the other, a process which would have to be accomplished by tunnelling, is vanishingly small. Prior to making an actual determination, however, there is no reason for supposing that the electron is associated with either proton rather than the other, since both protons are identical. The wave function for the electron must therefore reflect the fact that it is equally likely to be found associated with either proton.

If the wave function describing the system when the electron is associated with one proton is ψ_A, and that describing the system when the electron is associated with the other proton is ψ_B, the wave function describing the complete system prior to a determination of its actual configuration must contain both of these functions, suitably normalised to indicate the equal probability that an experiment determining the configuration will produce each of these results. The complete wave function for the molecular ion is thus

$$\psi = c_{\mathrm{I}}\psi_A + c_{\mathrm{II}}\psi_B \tag{7.1}$$

where c_{I} and c_{II} are suitable coefficients. It follows that

$$c_{\mathrm{I}}^2 \psi|_A|^2 = c_{\mathrm{II}}^2 |\psi_B|^2. \tag{7.2}$$

If $\psi_A = \psi_B$, that is, if the electron is considered to be in the same state on each proton, it follows that

$$c_{\mathrm{I}} = \pm c_{\mathrm{II}}. \tag{7.3}$$

There are thus two possible wave functions for the molecular ion corresponding to each state of the electron on the proton:

$$\psi_+ = N(\psi_A + \psi_B)$$

and

$$\psi_- = N(\psi_A - \psi_B). \tag{7.4}$$

Here, N is a normalising factor. The function ψ_+ is unchanged by a reversal of the two protons: that is, it is symmetrical with respect to the reversal of the proton coordinates. The function ψ_- reverses sign on an interchange of the two protons: it is antisymmetrical with respect to an interchange of the proton coordinates. It should be noted that although these solutions are referred to as "symmetrical" and "antisymmetrical" wave functions, this is *not* the same as the symmetry discussed in Chapter 5, which was with respect to an interchange of electron coordinates in a many-electron wave function: both the functions (7.4) are one-electron wave functions, and the symmetry is with respect to the coordinates of the centres of the two coulomb fields. At large separations both the wave functions ψ_A and ψ_B will be hydrogenic, but as the inter-proton distance diminishes, the increasing interaction of the two fields will modify the wave functions. The actual problem is very difficult, but some useful conclusions can be drawn from a qualitative analysis of a simple approximate field. In Fig. 7.1 the two fields are approximated by two square-well potentials, and the problem is further simplified by considering one dimension only. The wave functions for the two lowest states of this system are sketched in Fig. 7.2. The probability densities for these two distributions are very similar for large R; as shown in Fig. 7.3. However, as R decreases so that the overlap between the functions increases, the probability distributions for the symmetrical and antisymmetrical functions

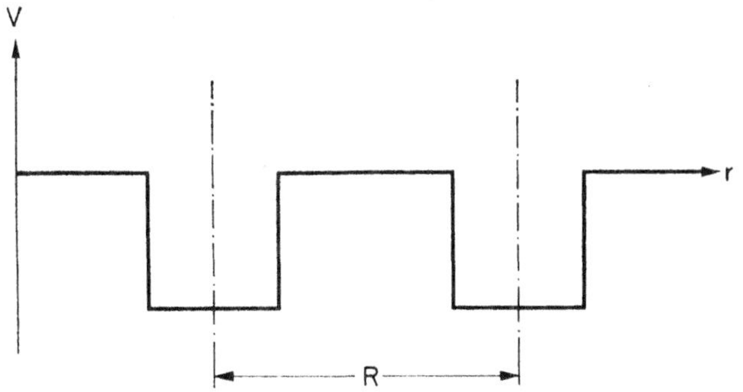

FIG. 7.1. Two square-well potential.

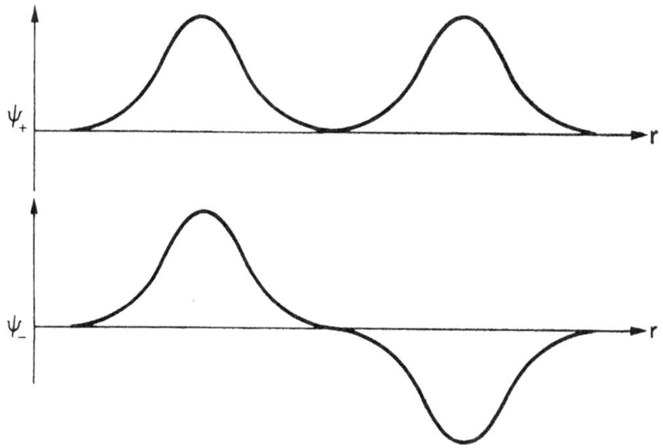

FIG. 7.2. The wave functions (7.4).

begin to show differences. The antisymmetrical function must always be zero midway between the protons, whereas the symmetrical function will be the sum of the two component functions at this point. At considerable interproton distances there is no significant difference, since both functions are effectively zero at this point anyway. The probability distributions are shown

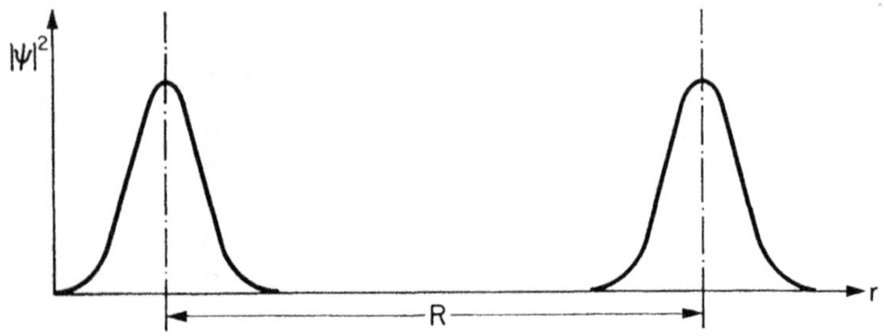

FIG. 7.3. The probability distribution for large R.

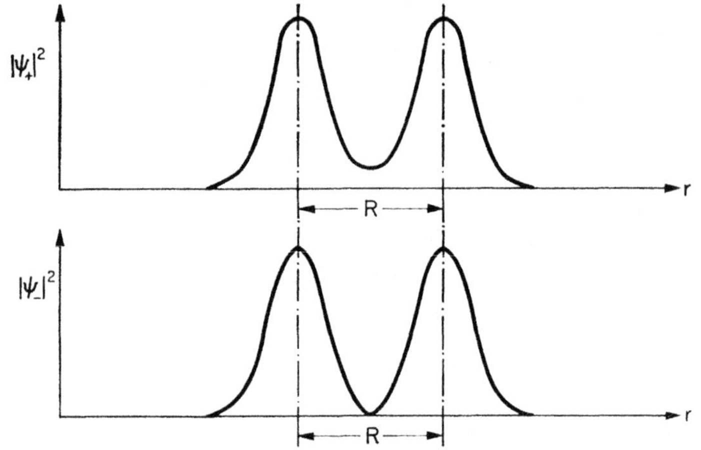

FIG. 7.4. Probability distributions for small R.

for small R in Fig. 7.4. It is plain that the energies of these two solutions are different. When the separation R is equal to the well width, the barrier between the wells vanishes, and the potential is a single well of twice the width. The symmetrical solution plainly approaches the ground state for this well, and the antisymmetrical solution, with a single node in the centre of the well, approaches the first excited level. Equation (5.3) gives the energy of a state in a square well with perfectly reflecting sides as

$$E_n = \frac{\hbar^2 k^2}{2ma^2} \cdot n^2$$

so that since the new well is twice the width of the first, the ground state energy is one-quarter that of the smaller well. The first excited state, for which $n = 2$, has an energy equal to the ground state energy of the smaller well. Although these values are only accurate for a well with perfectly reflecting sides, the general result, that the energy of the symmetrical solution will diminish as the distance between the wells decreases, finally falling to approximately one-quarter of the corresponding energy in the separate atoms. The energy of the antisymmetrical solution will vary only slightly with the separation between the two wells. The symmetrical function corresponds to binding between the two wells, consequently, whereas the antisymmetrical solution produces no binding. The form of the solutions drawn in Fig. 7.4 indicates that the electron spends an appreciable part of its time between the two wells for the symmetrical solution, and this may be regarded as representing the electron performing linking orbitals about the two protons. The general form of the energy of the two functions (7.4) as a function of R, the interproton distance, is sketched in Fig. 7.5. In fact, as shown, there is a slight repulsion corresponding to the antisymmetrical solution, as one would expect from the form of the solution sketched in Fig. 7.4. The total energy of the system includes the coulomb electrostatic repulsion between the protons, so the total energy of the molecular ion as a function of the interproton distance is shown in Fig. 7.6. The symmetrical function shows a minimum energy, so that a_0, the proton separation corresponding to the minimum energy, is the equilibrium separation of the stable molecular ion. The energy E_0, the minimum energy, is the binding energy of the

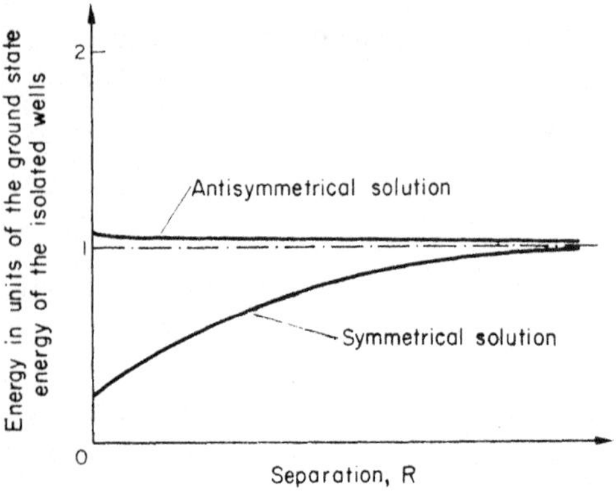

FIG. 7.5. The energies of the symmetrical and antisymmetrical solutions (7.4) as a function of well separation R.

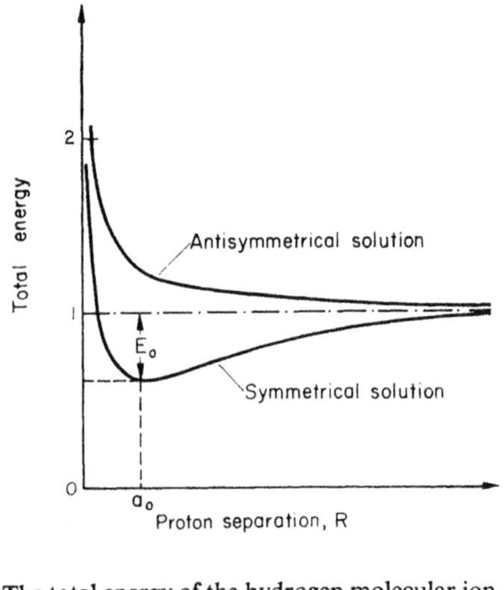

FIG. 7.6. The total energy of the hydrogen molecular ion, including Coulomb repulsion.

molecular ion. It is possible to calculate both these values from a wave mechanical model. The Schrödinger equation is

$$\nabla^2\psi + \frac{2m}{\hbar^2}\left(E + \frac{e^2}{r_a} + \frac{e^2}{r_b} - \frac{e^2}{R}\right)\psi = 0 \qquad (7.5)$$

where the distances r_a, r_b and R are shown in Fig. 7.7, in which A and B are the two protons and C is the electron. E_0 is calculated to be 63·9 kcal and a_0 to be 1·06 Å. The experimental binding

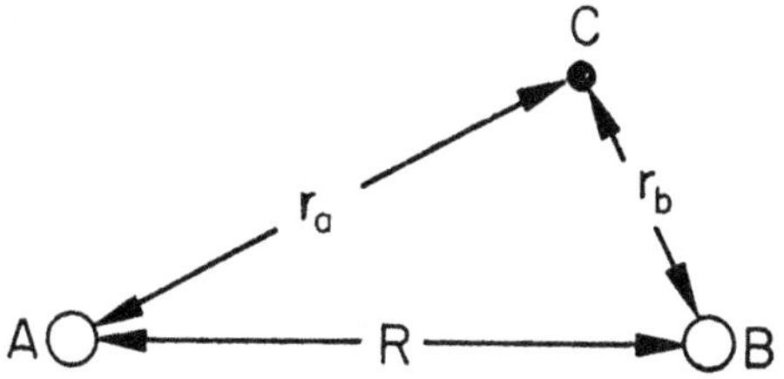

FIG. 7.7.

energy is 64·0 kcal and the observed internuclear spacing is 1·06 Å. The agreement between theory and experiment is very good.

It is easy to extend these qualitative arguments to more complex systems. Consider, for example, a potential consisting of three square-well potentials equally spaced in a line, a simplification of a hypothetical H_3^{++} ion. The wave function is

$$\psi = c_{\text{I}}\psi_A + c_{\text{II}}\psi_B + c_{\text{III}}\psi_C \qquad (7.6)$$

where as before the subscripts A, B and C label the protons, and the coefficients c_{I}, c_{II} and c_{III} are appropriate weighting constants. The Hamiltonian now contains several more terms, but is the same general form as (7.4). Again, since the chances of the electron being associated with any particular well are equal, it follows that so long as the component states are the same

$$c_{\text{I}}^2 = c_{\text{II}}^2 = c_{\text{III}}^2. \qquad (7.7)$$

There are consequently three possible wave functions:

$$\left.\begin{aligned} \psi_1 &\equiv c_{\rm I} = c_{\rm II} = c_{\rm III} \\ \psi_2 &\equiv c_{\rm I} = c_{\rm II} = -c_{\rm III} \\ \psi_3 &\equiv c_{\rm I} = -c_{\rm II} = c_{\rm III} \end{aligned}\right\} \quad (7.8)$$

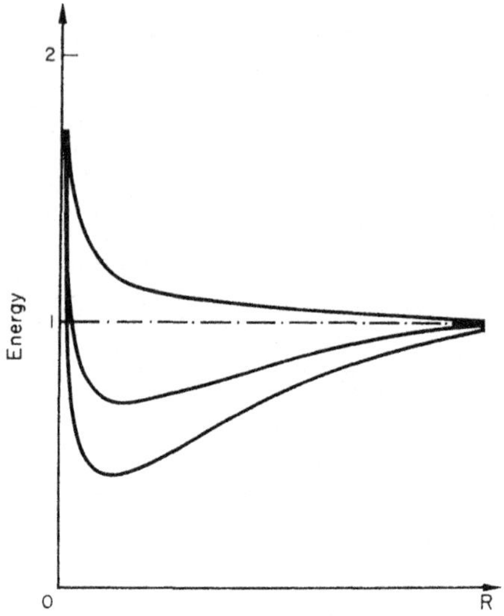

FIG. 7.8. The total energy of the three solutions (7.8).

all other possible combinations being equal to these, apart from a change of sign, or direction of numbering. In the limit, the first solution approaches the ground state of the large well formed when the barriers between them vanish: the second solution approaches the first excited state and the third solution approaches the second excited state. These have energies respectively one-ninth of, four-ninths of, and equal to the ground state energy of the separate atoms. Including the Coulomb repulsion term, the energies of the three states are sketched in Fig. 7.8. It is plain that these arguments can be extended to larger assemblies of atoms, and in general for an assembly of N atoms, each state of the free atom splits into N states in the assembly.

The Hydrogen Molecule H_2

This molecule is much more complicated, since the wave function is no longer a single-particle function, but must describe the motion and the interaction of two electrons. The potential energy V becomes

$$V = -\frac{e^2}{r_{A1}} - \frac{e^2}{r_{B1}} - \frac{e^2}{r_{A2}} - \frac{e^2}{r_{B2}} + \frac{e^2}{R_{AB}} + \frac{e^2}{r_{12}} \qquad (7.9)$$

where the distances are as shown in Fig. 7.9.

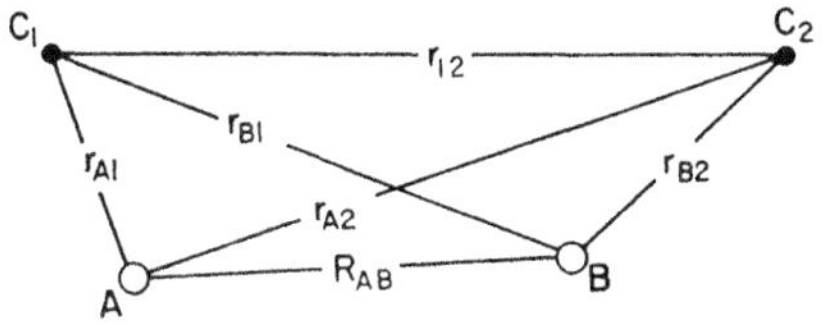

FIG. 7.9.

In line with the previous argument it may be assumed that at very large separations between the protons, the configuration of the system resembles two separate hydrogen atoms, with one electron associated with each proton. As before, the complete wave function for the system includes two functions, one the presenting the electron C_1 associated with proton A and re- electron C_2 associated with proton B; and one representing the reverse arrangement. These functions are labelled ψ_A and ψ_B respectively, and the complete wave function for the system is

$$\psi = c_{\text{I}}\psi_A + c_{\text{II}}\psi_B \qquad (7.10)$$

where again c_{I} and c_{II} are appropriate weighting functions. As before, either situation is equally likely, so that

$$c_{\text{I}} = \pm c_{\text{II}}. \qquad (7.11)$$

There are again therefore two solutions, each of which is analogous to the two solutions (7.4). However, both the component functions ψ_A and ψ_B are two electron functions. The Schrödinger equation for these solutions can be solved, and again the symmetrical solution gives an energy minimum. The value of the binding energy calculated from this minimum is 72·3 kcal, but the observed value is 108·8 kcal. It follows that the wave function (7.10) is incomplete. In fact there are two other possibilities: both electrons may be associated with either proton, leaving the other with no electron. Representing the wave functions for these two configurations by ψ_{AA} and ψ_{BB}, the complete wave function for the system may be written

$$\psi = c_{\text{I}}\psi_A + c_{\text{II}}\psi_B + c_{\text{III}}\psi_{AA} + c_{\text{IV}}\psi_{BB}. \qquad (7.12)$$

Since the two last configurations are equally likely, but are both much less likely than the first two, it follows that

$$c_{\text{I}} = \pm c_{\text{II}}$$

$$c_{\text{III}} = \pm c_{\text{IV}}$$

and

$$|c_{\text{I}}| > |c_{\text{III}}|. \qquad (7.13)$$

The wave function (7.12) gives a binding energy of 92·2 kcal, appreciably closer to the experimental value, although the agreement is still not very good. At large internuclear distances it is clear that the configurations described by the functions ψ_{AA} and ψ_{BB} are very unlikely, because of the electrostatic repulsion between the two electrons on the one proton, but at small distances between the protons this repulsion is compensated for by the coulomb attraction between the H^- ion with the two electrons, and the H^+ ion, the bare proton. The weighting factors c_{III} and c_{IV} are therefore functions of the distance between the protons; it is usual to say that the terms ψ_{AA} and ψ_{BB} are functions of the "overlap" between the wave functions of the separate atoms. Another factor which must be specifically included is the Pauli exclusion principle, or more properly, the symmetry requirements of the wave function (7.12) with respect to the co-ordinates of the two electrons. The states ψ_{AA} and ψ_{BB} are impossible if the two electrons have the same spin. On the other hand, there is an additional attraction if the electrons have opposite spin. The forces between the atoms are thus of three kinds: Coulomb attractions due to the differing charges on the atoms; overlap attractions due to the dependence of the co-efficients c_{III} and c_{IV} on the overlap between the wave functions; and exchange forces produced by the Pauli exclusion principle.

The two-electron bond produced between two like atoms by a pair of electrons with antiparallel spins is called the covalent bond. It is of course much stronger than the one-electron bond in the hydrogen molecular ion, in which exchange forces play no part. Since on average the electron spends equal time on each nucleus, the molecule has no resultant mean dipole moment, and for this reason the bond is also called a homopolar bond. At any given instant of time, however, the molecule will have a dipole moment, but the moment oscillates in such a way that the average dipole over a length of time is zero.

In constructing the wave function for the molecule (7.12), functions describing possible configurations of the system have been used. For those configurations for which the appropriate coefficients c_n are non-zero it is possible to interpret the result as a "resonance" of the system between the various configurations, the length of time that the molecule spends in any particular configuration being related to the magnitude of the coefficient c for that configuration. Although this concept is useful in visualising the problem, and has in fact been implicitly used in the preceding discussion, it must be understood that the philosophy of quantum mechanics does not really permit "quasi-particle" constructions of this kind.

Two-electron covalent bonds of this sort are formed between halogen atoms. The halogens all lack one electron from a full inert gas shell, and so the covalent bond has an additional stability because one of the possible configurations for each atom corresponds to a full shell. This is represented diagrammatically in Fig. 7.10, in which the open circles represent the outer electrons, and the full circles represent the rest of the atom. In Chapter 2 it was noted that as the atomic number goes up, the energy of an orbit depends increasingly on the eccentricity, because of the penetration of the electron into the atom core where it "sees" a higher charge. The increased attraction means that the electron spends more of its time in the vicinity of the nucleus and less time at the more remote parts of its orbit. In the free atom this produces an orbital precession and modifies

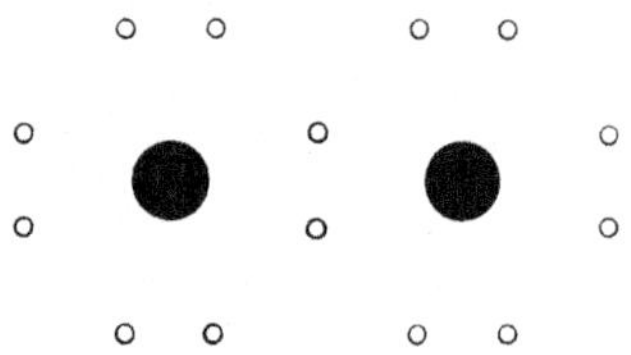

FIG. 7.10. A diagrammatic representation of a two-electron covalent bond between two halogen atoms.

the energy. However, in the diatomic molecule the "molecular orbits" performed by the electrons of the covalent bond as they link the two atoms will also penetrate into the atomic core, and as the atomic number goes up the electrons will spend an increasing fraction of their time near to the two nuclei, and a decreasing fraction between the two atoms. Consequently, the strength of the bond will decrease with increasing atomic number. This is demonstrated for the halogens in Table 7.1.

The elements of group VI in the Periodic Table lack two electrons from a full shell. They can achieve the desired inert gas electron structure by either (a) forming two electron covalent bonds with two other atoms, thus forming a chain or a ring. The most stable sulphur molecule is a puckered ring of eight atoms. Selenium also forms an eight-member ring in the vapour, and crystallises as long approximately linear chains, which are held together by van der Waals forces, which will be discussed later; or (b) forming a diatomic molecule with a double covalent bond involving four electrons, as shown in Fig. 7.11. This is the configuration of the stable O_2 molecule. S_2 molecules are detectable in sulphur vapour above 500°C, Se_2 molecules are

TABLE 7.1. BOND STRENGTHS FOR THE HALOGENS (FROM PAULING)

Bond	kcal/mol
F—F	63·5
Cl—Cl	57·8
Br—Br	46·1
I—I	36·2

present in the vapour phase, and Te_2 molecules are also present in tellurium vapour between 1400 and 1800°C. Again, the bond strength decreases with increasing atomic number: the double bond strength for oxygen is 117·2 kcal/mol, and for sulphur

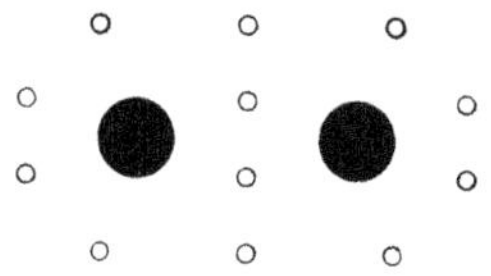

FIG. 7.11. A diagrammatic representation of a four-electron double covalent bond between two atoms from group VI.

102·6 kcal/mol. It can be seen that the bond strength is less than twice that of the equivalent single bond in Table 7.1; this arises in part because of the Coulomb repulsion of the electrons in the bond.

The elements of group V lack three electrons from a full shell; they may either form sheets of atoms, each linked to three others by single covalent bonds, or form treble covalent bonds involving six electrons between two atoms, as shown in Fig. 7.12. This is the configuration of the N_2 molecule, and the triple bond strength is 170 kcal/mol. P_2 molecules are also stable, with a bond strength of 116 kcal/mol; the bond strength is quite appreciably less than three times the single bond strength listed in Table 7.1. However, various other configurations are possible: arsenic and antimony form predominantly molecules containing four atoms, as does phosphorus in the vapour. In the solid, arsenic, antimony

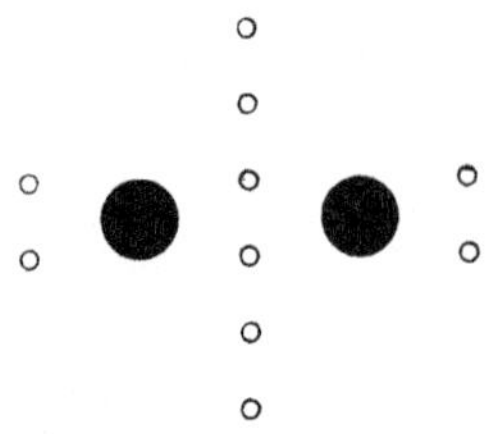

FIG. 7.12. A diagrammatic representation of a six-electron triple covalent bond between two atoms from group V.

and bismuth form large approximately coplanar sheets of atoms, each covalently linked to three other atoms. The sheets are held together in the solid by van der Waals forces. The nature of the elements of group V becomes noticeably more metallic as the atomic number increases: phosphorus is clearly a non-metal, but arsenic, antimony and bismuth have a number of metallic characteristics.

In group IV the atoms lack four electrons from a full shell. In the diamond structure adopted by one form of carbon and by silicon and germanium each atom is linked by single covalent bonds to four other atoms, forming a three dimensional array. The diamond structure is sketched in Fig. 7.13. This array can be extended without limit, so that each crystal is in reality a giant molecule. Since the bonding in the solid is purely covalent, the melting point is a measure of the bond strength: these are tabulated in Table 7.2, and it can be seen that the bond strength diminishes as the atomic number increases. Again, the character of the elements becomes increasingly more metallic as the atomic

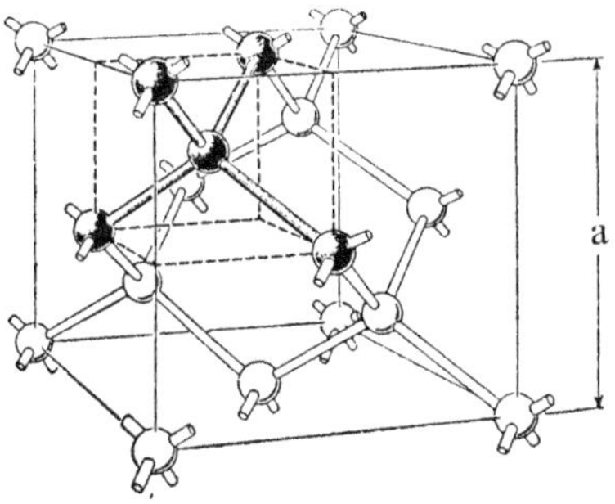

FIG. 7.13. The crystal structure of diamond. (From Shockley's *Electrons and Holes in Semiconductors*, Copyright 1950, D. Van Nostrand Company, Inc., Princeton, N.J.)

number increases: carbon is a non-metal; silicon and germanium are semiconductors with many metallic characteristics; tin exists in two modifications: the low temperature form, grey tin, is a

TABLE 7.2. THE MELTING POINTS OF THE ELEMENTS OF GROUP IV

Element	Melting point (°C)
Carbon	3700
Silicon	1415
Germanium	958
Grey tin	—
White tin	232
Lead	327

semiconductor, and the high temperature form, white tin, is a true metal. Lead, the final member in the group, is also a true metal.

The reason for the abrupt change in character at the end of the table arises because the four electrons in the free atom consist of two *s* electrons and two *p* electrons. The *s* state is thus full, and consequently only two electrons—those in the *p* state—are available to form bonds. However, in the elements of lower atomic number the *s* and *p* states are of very similar energy, and they combine to form four hybrid (*sp*) states, each containing one electron, all available to form bonds. This is called "promotion". As the atomic number increases the energy difference

TABLE 7.3. CARBON–CARBON BOND STRENGTHS (AFTER PAULING)

Bond	Bond energy (kcal/mol)
C—C	58·6
C = C	100
C ≡ C	123

between the *s* and *p* states increases and promotion becomes increasingly difficult: the hybrid states have low stability for tin and do not form at all for lead.

For white tin and lead, and for the elements of groups I, II and III, there are insufficient electrons for the atom to form a stable inert gas configuration by sharing electrons, as in the covalent bond. However, it is possible to envisage an arrangement in which a bonding pair of electrons "resonates" between a number of atoms, so that for a part of the time each atom has a stable shell. This model implies a degree of mobility in the valence electrons, which is not possessed by the electrons in the non-resonating type of covalent bond considered before. This degree of mobility of the valence electrons suggests that these elements will conduct electricity, whereas solids like diamond in which the electrons are immobile will be insulators. This is in fact the case: the elements of groups I, II and III together with white tin and lead are in fact metals. This description of metallic bonding in terms of resonating covalent bonds has been developed by Pauling.

Covalent bonding is also important between unlike atoms, and in fact the large and important class of organic molecules are largely bonded by covalent bonds. Carbon atoms in these molecules may be linked by single, double or treble covalent bonds: the bond strengths are tabulated in Table 7.3, and it can be seen again that the bond strength increases, but not regularly, because of the mutual repulsion of the electrons.

Directionality of Bonding

Analysis of the charge distribution around an atom due to the electrons shows that a full shell always has spherical symmetry. A half-full shell has its lowest energy when there is one electron in each doubly degenerate state defined by the first three quantum numbers—this is called Hund's rule—and this, too, has spherical symmetry. It follows that *s* states which only accommodate two electrons always have spherical symmetry. Consequently, covalent bonds formed from *s* electrons are not directional. This spherical symmetry can be deduced from the wave function, which for *s* states is a function of *r* only:

$$\psi = f(r). \tag{7.14}$$

The three *p* states however are not spherically symmetrical. In fact,

$$\left.\begin{aligned} \psi_{m=0} &= \frac{x}{r} f(r) \\ \psi_{m=-1} &= \frac{y}{r} f(r) \\ \psi_{m=+1} &= \frac{z}{r} f(r) \end{aligned}\right\} \tag{7.15}$$

so that the three *p* states are mutually perpendicular (in classical terms, the major axes of the ellipses corresponding to the three orbits would be at right angles), and covalent bonds formed from *p* states will also tend to be at right angles to each other. The four (*sp*) hybrid states formed by the group IV elements are at an angle of 109° 28′. This is the angle between the lines joining the centre of gravity of a regular tetrahedron to its apices, and accounts for the tetrahedral arrangement of atoms in the diamond structure. The strong directionality of the bonds in covalently linked materials accounts for their high rigidity, whereas the resonating bonds in the metals will plainly not have any marked directionality, accounting for the ductility of metals.

Forces Between Neutral Atoms and Molecules

Atoms of the inert gases, and molecules which have achieved inert gas electronic structures have no long-range attractive Coulomb forces. There is a non-directional repulsion due to overlap of the electron clouds at short range, and the magnitude of the repulsion increases rapidly as the distance decreases. However, although the mean dipole moment of a neutral atom is zero, as also are the moments of higher order, the mean square dipole moment is not zero. An external dipole can induce a dipole moment in a neutral atom, and thus there is a tendency for the oscillating dipole moments of two neutral atoms to couple, producing a net attraction.

Consider two chemically identical and electrically neutral molecules, the centres of which are kept at a constant distance r apart. Suppose that each molecule possesses one electron which vibrates harmonically in the direction of the line of centres. If at any instant the electron in the first molecule is displaced by a distance x_1 from the massive nucleus which will be approximately at rest, it acquires a displacement energy

$$u_1 = \tfrac{1}{2} f x_1^2 \tag{7.16}$$

where f, the restoring constant, is e^2/α, where α is the polarisability of the molecule. At the same instant, the displacement of the electron in the second molecule is x_2, and the displacement energy is u_2, where again

$$u_2 = \tfrac{1}{2} f x_2^2. \tag{7.17}$$

The frequency of vibration of the two molecules is the same, so long as the oscillations may be regarded as independent:

$$\nu = \frac{1}{2\pi}\sqrt{\left(\frac{f}{m}\right)} = \frac{1}{2\pi}\sqrt{\left(\frac{e^2}{m\alpha}\right)}. \tag{7.18}$$

In addition, the energy of the system contains an electrostatic interaction term:

$$u_3 = \frac{e^2}{r} + \frac{e^2}{r+x_1+x_2} - \frac{e^2}{r+x_1} - \frac{e^2}{r+x_2}. \tag{7.19}$$

If x is small compared to r, each term may be expanded by means of the binomial theorem. The second term gives

$$e^2\left\{\frac{1}{r} - \frac{x_1+x_2}{r^2} + \frac{(x_1+x_2)^2}{r^3} - \ldots\right\} \tag{7.20}$$

the third

$$e^2\left\{\frac{1}{r} - \frac{x_1}{r^2} + \frac{x_1^2}{r^3} - \ldots\right\} \tag{7.21}$$

and the last

$$e^2\left\{\frac{1}{r} - \frac{x_2}{r^2} + \frac{x_2^2}{r^3} - \ldots\right\}. \tag{7.22}$$

The first non-vanishing term in the expansion of (7.19) is thus

$$\frac{2e^2 x_1 x_2}{r^3} \tag{7.23}$$

and since it has been assumed that both x are very much smaller than r, later terms in the expansion can be neglected in comparison to the first. The electrostatic interaction energy is thus

$$u_3 \cong \frac{2e^2 x_1 x_2}{r^3}. \tag{7.24}$$

The total potential energy V is then

$$V = u_1 + u_2 + u_3 = \tfrac{1}{2}f(x_1^2 + x_2^2) + \frac{2e^2 x_1 x_2}{r^3}. \quad (7.25)$$

Now set

$$x_s = \frac{1}{\sqrt{2}}(x_1 + x_2); \qquad x_a = \frac{1}{\sqrt{2}}(x_1 - x_2)$$

and writing

$$f_s = f\left(1 + \frac{2\alpha}{r^3}\right); \qquad f_a = f\left(1 - \frac{2\alpha}{r^3}\right)$$

the potential energy becomes

$$V = \tfrac{1}{2}f_s x_s^2 + \tfrac{1}{2}f_a x_a^2. \quad (7.26)$$

In terms of these variables the Schrödinger equation for the system may be written

$$\frac{\partial^2 \psi}{\partial x_s^2} + \frac{\partial^2 \psi}{\partial x_a^2} + \frac{2m}{\hbar^2}(E - \tfrac{1}{2}f_s x_s^2 - \tfrac{1}{2}f_a x_a^2)\psi = 0. \quad (7.27)$$

The energy eigenvalue spectrum is

$$E = (n_s + \tfrac{1}{2})h\nu_s + (n_a + \tfrac{1}{2})h\nu_a \quad (7.28)$$

where n_s and n_a are integers. The lowest eigenvalue is that for which both n_s and n_a are zero:

$$E_0 = \tfrac{1}{2}h(\nu_s + \nu_a). \quad (7.29)$$

The lowest energy of a system of two independent oscillators is

$$E_0' = 2(\tfrac{1}{2}h\nu) \quad (7.30)$$

where here ν is the frequency of the independent oscillators (the same for both). The gain in the ground state energy due to the coupling is thus

$$\Delta E_0 = E_0 - E_0' = \tfrac{1}{2}h(\nu_s + \nu_a) - h\nu. \quad (7.31)$$

Substituting for ν_s and ν_a,

$$\Delta E_0 = \tfrac{1}{2}h\nu\left[\left(1+\frac{2\alpha}{r^3}\right)^{\frac{1}{2}}+\left(1-\frac{2\alpha}{r^3}\right)^{\frac{1}{2}}-2\right]. \tag{7.32}$$

Expanding, and neglecting powers of $2\alpha/r^3$ beyond the second, we obtain

$$\Delta E_0 \cong -\frac{h\nu\alpha^2}{2r^6}. \tag{7.33}$$

This result signifies that between two neutral molecules or atoms behaving as oscillating dipoles there is a net attraction which varies inversely as the sixth power of their separation. This conclusion is due to London (1930). The above treatment is simplified by the assumption of a one-dimensional form for the equations of motion, but the alteration is slight for an extension to three dimensions: in this case the change in the ground state energy due to the interaction is

$$\Delta E_0 = -\frac{3}{4}\cdot\frac{h\nu\alpha^2}{r^6}. \tag{7.34}$$

The forces produced by this attraction are called van der Waals forces, and they are responsible for the bonding in solid and liquid inert gases; for the bonding between molecules in the liquid and solid halogens; for the bonding between the ring or line molecules in liquid and solid group VI elements and the sheet molecules in group V elements. Since the attraction depends on the dipole induced in a molecule, one would anticipate that the magnitude of the attraction would increase as the atomic number increases: this suggests that the melting points of the elements in the last four groups of the Periodic Table should increase with increasing atomic number. The melting points are listed in Table 7.4.

The variation of melting point is in accord with the prediction, except for bismuth, in which the metallic characteristics are well developed. It should be also noticed that as the size of the molecular unit bonded by the van der Waals forces increases

TABLE 7.4. THE MELTING POINTS OF THE ELEMENTS OF THE LAST FOUR GROUPS IN THE PERIODIC TABLE

Group V element	M.P. (°C)	Group VI element	M.P. (°C)	Group VII element	M.P. (°C)	Group VIII element	M.P. (°C)
						He	−271·4
N	−210·0	O	−218·8	F	−223	Ne	−248·6
P	44·1	S	112·8	Cl	−101	Ar	−189·4
As	220	Se	220	Br	−7·2	Kr	−157
Sb	450	Te	450	I	114	Xe	−112
Bi	271·3	Po	—	At	—	Rn	−71

(from right to left in Table 7.4) the melting point again increases, apart from slight irregularities as the bonding becomes more complex in group V. While it is true to say that the van der Waals forces are relatively weak, they are not so weak that they can be disregarded. Equations (7.33) and (7.34) demonstrate that the van der Waals forces are non-directional in character.

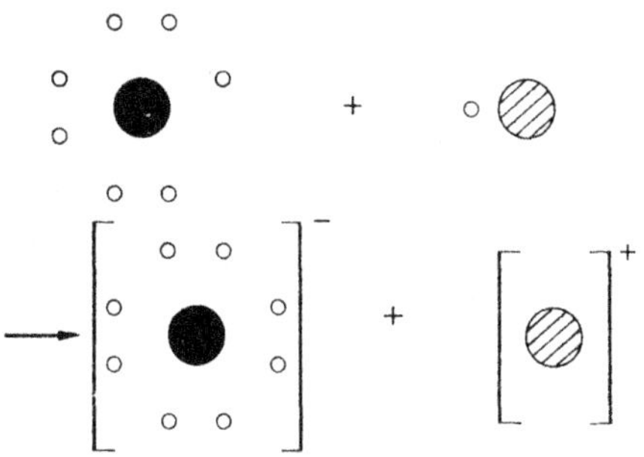

FIG. 7.14. Diagrammatic representation of electron transfer in the reaction between a halogen (shaded) and an alkali metal (cross-hatched).

Bonding Between Unlike Atoms

The forces between unlike atoms are similar to those between like atoms. However, the symmetry considerations which led to the relative values (7.13) ascribed to the coefficients in the wave function (7.12) no longer apply if the two atoms are different. Consider, for example, the case of a sodium atom and a chlorine atom. Since the chlorine atom lacks one electron from a full shell, and the sodium atom has one electron outside a full shell, the molecular configuration in which the chlorine atom gains one electron from the sodium atom is very much more stable than the other possible configurations, and the weighting coefficient c_i for this arrangement will therefore be much greater than any other. In effect, therefore, the electron may be regarded as completely transferred, as symbolised in Fig. 7.14. Now, the exchange forces and the overlap forces which form such an important part of the covalent bonding attractions are of minor importance: the principal force between the two atoms is the coulomb attraction arising from their different charges. The sodium atom having lost an electron has the electronic configuration of the inert gas neon, and a net positive charge: it is now called a sodium ion. The chlorine atom having gained an electron has the electronic structure of the inert gas argon, and a net negative charge: it is called a chlorine ion. In general, positively charged ions are called cations and negatively charged ions are called anions. The bonding between two unlike atoms which arises from the effective complete transfer of an electron is called ionic bonding: it is coulombic in origin, and is therefore non-directional. It should be noted that there is really no fundamental difference between a covalent and an ionic bond: the difference merely lies in the degree of importance of the various contributions to the bond energy. The homopolar bond formed between two similar atoms and the ionic bond formed between an alkali metal and a halogen are the two extremes of a continuous spectrum of bond types. In general, a two-electron bond formed between two unlike atoms will be described by a wave function similar to (7.12), but none of the conditions (7.13) will necessarily hold. In particular, if c_{III} is not equal to c_{IV}, the bond will have a net dipole, and there will be a relatively large coulomb contribution to the bonding. Sometimes such a bond is described as being partly covalent and partly ionic, and in quantitative terms the magnitude of the dipole moment may be quoted as a percentage of the comparable wholly ionic dipole. In general, a two-electron bond which has a net permanent dipole is called a heteropolar bond, in contrast to the homopolar bond formed between two like atoms with no net permanent dipole.

Consider the two hydrogen–oxygen bonds in the water molecule. In this case, both electrons in each two-electron bond tend to remain in the vicinity of the oxygen atom, so that the bond is heteropolar; the oxygen atom has a net negative charge, and each hydrogen atom has a net positive charge. The bonding orbitals are the $2p$ orbitals of the oxygen atom, so that in accord with (7.15), the angle between the two bonds should be 90°. However, the coulomb repulsion between the two positively charged hydrogen atoms tends to dilate the bond angle: experimentally the bond angle is determined to be 105°. In the similar F_2O molecule, however, the bond is less polar, and the bond angle is only 100°; and in H_2S the bonds are very nearly homopolar, and the bond angle is 92° 20′.

The Hydrogen Bridge

A further type of bond of considerable importance is the hydrogen bond or "bridge". A hydrogen atom may under some circumstances form bonds with *two* atoms, rather than only one as would normally be expected. This hydrogen bond is quite weak, having a bond energy of only 5 kcal/mol; but it is responsible for such important processes as the polymerisation in water and in hydrogen fluoride; and is also important in certain ferroelectric crystals, notably potassium dihydrogen phosphate. The hydrogen bond may be qualitatively visualised as follows: if the normal two-electron bond between a hydrogen atom and another atom is sufficiently heteropolar, the hydrogen atom may be regarded as a bare proton, attracted to the negative ion and forming a stable pair with it at the characteristic bond distance. The proton may then be thought of as a point cation, the other atom as an anion sphere of finite radius. The positive charge on the proton is capable of attracting other anions, but when another has been attracted the two anion spheres are in contact, inhibiting the attraction of other anions to the proton. It follows that only molecules in which the normal hydrogen bond is noticeably heteropolar can form a hydrogen bridge.

The Metallic Bond

No summary of bonding between atoms would be complete without a reference to metallic bonding, but unfortunately this bond is very difficult to describe in the semi-qualitative terms that have been used for the other types of bond listed above. The simplest concept of the metallic bond is that of the resonating covalent bonds, developed by Pauling, which explains the conductivity of metals and the non-directional nature of the bonding which is responsible for their high ductility. The model is, however, of little value for analytical purposes. A complete analysis is one of the more advanced problems in quantum mechanics, and is beyond the scope of the present book.

CHAPTER 8

Atoms in Motion

THE forces between atoms and molecules which are electrostatic in origin, such as the coulomb interaction of two charges, the van der Waals attraction of two oscillating dipoles, the interaction between an oscillating dipole and a permanent dipole, and the interaction between two molecules each having a permanent dipole, all vary as an inverse power of the separation. All these attractive forces may be represented by a potential field of the form

$$u = -Br^{-m} \tag{8.1}$$

where m and B are constants, depending on the nature of the two molecules involved, and r is their separation. Although (8.1) is spherically symmetrical, and in fact all the attractions listed above are also spherically symmetrical, angular dependence of u may be included by allowing B to be a function of angle. Equation (8.1) does not, of course, necessarily apply to attractions due to exchange, and repulsive forces apart from the coulomb repulsion of two like charges are also difficult to analyse. However, in 1903 Mie suggested that in some cases the repulsive forces might be described by a similar function, and that the total potential energy of a pair of atoms or molecules could be represented by a function of the form

$$u_{\text{total}} = u_{\text{repulsion}} + u_{\text{attraction}} = Ar^{-n} - Br^{-m} \tag{8.2}$$

where A is a positive constant, and $n > m$.

E*

The equilibrium separation r_{eq} of a pair of molecules subject to this potential may be determined by minimising the energy:

$$-\left(\frac{du}{dr}\right)_{r=r_{eq}} = 0. \tag{8.3}$$

Hence,

$$r_{eq}^{n-m} = \frac{nA}{mB}. \tag{8.4}$$

The equilibrium energy u_{eq} is then

$$u_{eq} = Br_{eq}^{-m}\left(\frac{m}{n}-1\right) = Ar_{eq}^{-n}\left(1-\frac{n}{m}\right). \tag{8.5}$$

Eliminating A and B, Mie's equation (8.2) becomes

$$u = u_{eq}\frac{1}{(m-n)}\left[m\left(\frac{r_{eq}}{r}\right)^{n} - n\left(\frac{r_{eq}}{r}\right)^{m}\right]. \tag{8.6}$$

The dissociation energy of the stable pair is the energy required to increase the separation between them from $r = r_{eq}$ to $r = \infty$: that is,

$$D_e = u_{r=\infty} - u_{r=r_{eq}} = -u_{eq}. \tag{8.7}$$

The excess energy at a separation r over that at a separation r_{eq} is

$$w = u - u_{eq} = \frac{D_e}{n-m}\left\{n\left[1-\left(\frac{r_{eq}}{r}\right)^{m}\right] - m\left[1-\left(\frac{r_{eq}}{r}\right)^{n}\right]\right\}. \tag{8.8}$$

Now consider how this excess energy w varies with small displacements of the bond length from the equilibrium value. Rewriting (8.8) in terms of the variable

$$\left.\begin{aligned} x &= \frac{r-r_{eq}}{r_{eq}} \\ w &= \frac{D_e}{n-m}\{n[1-(1+x)^{-m}] - m[1-(1+x)^{-n}]\}. \end{aligned}\right\} \tag{8.9}$$

Expanding the terms in brackets and neglecting powers of x greater than the second,

$$w = \tfrac{1}{2}mnD_e x^2 = \tfrac{1}{2}mnD_e \left(\frac{r-r_{eq}}{r_{eq}}\right)^2. \tag{8.10}$$

The displacement energy is thus proportional to the square of the displacement. This means that the pair of molecules will execute simple harmonic motion about the equilibrium separation, with a frequency

$$\nu_{eq} = \frac{1}{2\pi r_{eq}}\sqrt{\left(\frac{mnD_e}{\mu}\right)} \tag{8.11}$$

where μ is the reduced mass of the system.

Note that for small displacements the terms in brackets in (8.9) may be rewritten

$$\left.\begin{aligned}(1+x)^{-m} &\to \exp(-mx)\\ (1+x)^{-n} &\to \exp(-nx)\end{aligned}\right\} \tag{8.12}$$

so that (8.8) becomes

$$w = \frac{D_e}{n-m}\{n[1-\exp(-mx)]-m[1-\exp(-nx)]\}. \tag{8.13}$$

This expression also when expanded gives (8.10), and as $x \to \infty$, $w \to D_e$. Equation (8.13) is thus an acceptable alternative to Mie's equation. In particular, if $n = 2m$ and $m = Kr_{eq}$, (8.13) becomes

$$w = D_e\{1-\exp[-K(r-r_{eq})]\}^2 \tag{8.14}$$

which is the same as an equation suggested on empirical grounds by Morse in 1929 to describe the energy of a pair of atoms in a diatomic molecule. The choice of a potential function is therefore somewhat arbitrary, and a large number of functions similar to the two given above have been used by various investigators. All share the property that for small displacements from the equilibrium, the bond length varies with simple harmonic motion.

As the displacement increases terms of higher order become important: these are referred to as anharmonic terms. When the potential (8.14) is substituted into Schrödinger's equation the energy eigenvalue spectrum is

$$E = (n+\tfrac{1}{2})\hbar\omega - \frac{(\hbar\omega)^2}{4D_e}(n+\tfrac{1}{2})^2 \qquad (8.15)$$

where n is an integer.
Provided $(n+\frac{1}{2})\hbar\omega \ll D_e$ this does not differ significantly from the spectrum for the simple harmonic oscillator (6.25).

The Potential Energy of a System of Molecules

In an assembly of molecules subject to the type of forces described by Mie's equation, each molecule may be regarded as forming a "pair" of the type described above with every other molecule in the assembly. If there are N molecules in the assembly, each molecule is therefore involved in $(N-1)$ pairs: the total number of pairs formed is thus $\frac{1}{2}N(N-1)$. The factor $\frac{1}{2}$ appears in this expression because each pair involves two molecules. If N is very large, the number of pairs is effectively $\frac{1}{2}N^2$. However, because of the rapid decrease in the magnitude of the interactions between molecules as the distance between them increases, it is usually sufficient to consider only the interactions between a molecule and its nearest neighbours. Suppose each molecule in the assembly has c nearest neighbours, the number of "nearest-neighbour pairs" in the assembly is then $\frac{1}{2}Nc$. c is called the co-ordination number of the molecular arrangement. If the average energy of a single pair is $\bar{E}$, the potential energy of the complete assembly is

$$U = \tfrac{1}{2}Nc\bar{E}. \qquad (8.16)$$

Mie's equation (8.2) gives $\bar{E}$, so (8.16) becomes

$$U = \tfrac{1}{2}Nc(Ar^{-n} - Br^{-m}). \qquad (8.17)$$

The molecules outside the first coordination sphere may be

included by incorporating two constants s_n and s_m, which, if the nearest-neighbour approximation is valid, will not greatly exceed unity:

$$U = \tfrac{1}{2}Nc(s_n A r^{-n} - s_m B r^{-m}). \tag{8.18}$$

The evaluation of the constants s_n, s_m is largely a geometric problem.

The average potential energy of a single molecule in the assembly is then

$$u = U/N = \tfrac{1}{2}c(s_n A r^{-n} - s_m B r^{-m}). \tag{8.19}$$

The condition for equilibrium is

$$-\left(\frac{du}{dr}\right)_{r=r_s} = 0 \tag{8.20}$$

where r_s is the nearest-neighbour distance corresponding to minimum potential energy. Applying this condition to (8.19) gives

$$r_s^{n-m} = \frac{s_n n A}{s_m m B} \tag{8.21}$$

and the minimum potential energy is

$$u_s = \tfrac{1}{2}cs_m B r_s^{-m}\left(\frac{m}{n}-1\right) = \tfrac{1}{2}cs_n A r_s^{-n}\left(1-\frac{n}{m}\right). \tag{8.22}$$

Eliminating A and B from (8.19) and (8.22), the potential energy at an arbitrary nearest-neighbour distance r is

$$u = \frac{u_s}{m-n}\left[m\left(\frac{r_s}{r}\right)^n - n\left(\frac{r_s}{r}\right)^m\right]. \tag{8.23}$$

Now, the energy required to increase the nearest-neighbour separation from the equilibrium value r_s to infinity is plainly the energy required to vaporise the assembly: this is called the internal heat of sublimation if the assembly is a solid, and of vaporisation if the assembly is a liquid. It is plainly equal to $-u_s$. It is clear from the similarity of (8.23) to (8.6) that an atom in an assembly will vibrate with simple harmonic motion about its mean position so long as the displacements are small, but that for larger displacements there will be in addition anharmonic terms in the motion.

The coefficients n and m in Mie's equation are determined by the interaction between atoms and so may be experimentally determined by studying properties of molecular assemblies which depend on the molecular interactions. For example, the ideal gas laws are derived assuming no molecular interaction. Consequently, any deviation from ideality in the behaviour of real gases may be used as a measure of the intermolecular forces. For solids, the compressibility is plainly determined by the form of (8.23), and in fact the product mn may be obtained from the compressibility. The repulsion constant n varies from values as low as 3 for C_6H_6 to 13 for Ne. The higher the value of n, the closer the molecule is to a rigid solid, so that remembering the spherical symmetry of the inert gas electronic structure, it is reasonable to regard neon atoms as rigid spheres. For neutral atoms and molecules the attractive coefficient m has the value 6, since the attractive forces are the van der Waals forces.

It is interesting to note that in KCL both ions have the same electronic structure as argon, and that although in one case the attractive forces are due to the coulomb attraction and in the other to the van der Waals forces, the forces of repulsion, which arise from overlap of the electron shells, should be the same in each case. Experiment shows that for KCl

$$u_{\text{repulsion}} = Ar^{-n} = 6{\cdot}55 \times 10^{-81} \text{ erg cm r}^{-9}$$

and for argon

$$u_{\text{repulsion}} = 6{\cdot}02 \times 10^{-81} \text{ r}^{-9}.$$

The similarity of these two results confirms the usefulness of the concepts of Mie and similar potentials to describe the interactions of atoms and molecules in solids and liquids in systems where the forces can be regarded as purely centrally directed. For metals, however, potentials like Mie's are of limited value. Recently, some slight improvement in the description of metals has been obtained by use of a potential originally suggested by Rydberg:

$$u = -D_e[1-b(r-r_e)]\exp[-b(r-r_e)]\,. \tag{8.24}$$

Propagation of an Elastic Wave

Consider a homogeneous elastic line, and suppose that the motion of each element of the line is constrained to be parallel to the line itself. Let x be the position coordinate of an element of the line, and let u be the displacement of the element from its equilibrium position. The strain is defined as the fractional change in length:

$$\varepsilon = \frac{\partial u}{\partial x}. \tag{8.25}$$

If the force producing this strain is F, then Hooke's law states that

$$F = c\varepsilon \tag{8.26}$$

where the constant of proportionality c is called the elastic stiffness constant. Suppose further that the line density, defined as the mass per unit length, is ρ. Now consider an element of the line length Δx. The strain at one end is $\varepsilon(x)$ and at the other $\varepsilon(x+\Delta x)$, and

$$\begin{aligned}\varepsilon(x+\Delta x) &= \varepsilon(x)+\frac{\partial \varepsilon}{\partial x}\Delta x\\ &= \varepsilon(x)+\frac{\partial^2 u}{\partial x^2}\Delta x.\end{aligned} \tag{8.27}$$

The resultant force on the element is thus

$$\left[\varepsilon(x)+\frac{\partial^2 u}{\partial x^2}\Delta x-\varepsilon(x)\right]c = c\frac{\partial^2 u}{\partial x^2}\Delta x. \tag{8.28}$$

From Newton's second law the acceleration of the element is $\partial^2 u/\partial t^2$, where

$$c\frac{\partial^2 u}{\partial x^2}\Delta x = \rho\Delta x\frac{\partial^2 u}{\partial t^2}. \tag{8.29}$$

So the equation of motion of the line is

$$\frac{\partial^2 u}{\partial x^2} = \left(\frac{\rho}{c}\right)\frac{\partial^2 u}{\partial t^2}. \tag{8.30}$$

The most general solution of this equation is

$$u = F(x-vt)+G(x+vt) \tag{8.31}$$

where the first part of the solution corresponds to a disturbance, whose shape initially is $F(x)$, where F represents any function at all, moving from left to right with a velocity v where

$$v = \left(\frac{c}{\rho}\right)^{\frac{1}{2}} \tag{8.32}$$

without altering its shape; and the second part of the solution describes a disturbance whose initial shape is $G(x)$, where again G is any function at all, travelling at the same velocity in the opposite direction, again without change of shape. The accuracy of this general solution may be confirmed by direct substitution in (8.30).

It should be noted that the velocity of propagation is a property only of the medium, not of the form of the disturbance, and that propagation of a disturbance along the line takes place without change of shape. This is called non-dispersive propagation.

Since (8.26) implies that the force on an element of the line is proportional to its displacement from its equilibrium position, it follows that each element of the line will vibrate with simple harmonic motion. For this reason, it is usual to select from the generality of functions which satisfy (8.30), functions which have wave form. These may be sine or cosine waves, but it is generally more convenient to use functions of the form

$$u = \exp i(\omega t \pm kx). \qquad (8.33)$$

This wave equation and its properties have already been discussed briefly in Chapter 4.

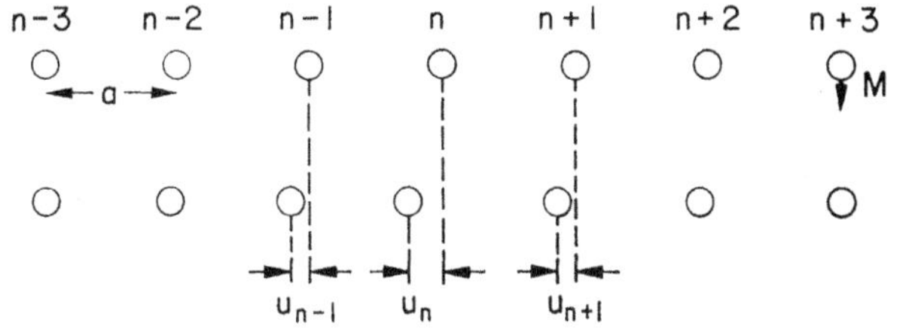

FIG. 8.1. Propagation of an elastic wave along a line of similar atoms.

Wave Motion of a Line of Similar Atoms

In fact, a real solid is not an elastic continuum, but consists of discrete atoms. The macroscopic properties of a solid are not dissimilar to those of an elastic continuum because the atoms are so small, but it is to be expected that when the wavelength of an elastic wave becomes comparable with the interatomic spacing, the propagation may differ from that of a continuum. In one dimension, the problem may be solved relatively simply. The atom positions at equilibrium are shown in the upper part of Fig. 8.1. Suppose each atom has mass M, and the spacing of the atoms along the line—the lattice parameter—is a. In the first instance it will be assumed that the line is of infinite length, and the atoms are numbered from the left $1, 2, \ldots, (n-1), n, (n+1), \ldots$ etc., from some arbitrary point. Now consider the motion of the nth atom after being released from a position different from its equilibrium position, assuming nearest neighbour interactions only to be important. As was clear from the analysis of Mie's equation, the general solution will differ from the nearest neighbour solution only by two geometric terms. Furthermore, it will be assumed that the displacements are small, so that only harmonic interactions need be included.

Let the nth atom be displaced a distance u_n from its equilibrium position in the direction of the line, and released at time zero. This situation is represented in the lower part of Fig. 8.1. The force acting on the nth atom is then F_n, where

$$F_n = \beta(u_{n+1}-u_n)-\beta(u_n-u_{n-1}). \tag{8.34}$$

The first term on the right-hand side represents the increase in length of the bond from atom n to atom $n+1$, the second represents the increase in the bond length from atom n to atom $n-1$. β is called the force constant, and is analogous to the elastic stiffness constant in the elastic continuum case. The equation of motion of the nth atom is thus

$$M\frac{\partial^2 u}{\partial t^2} = \beta(u_{n+1}+u_{n-1}-2u_n). \tag{8.35}$$

In the long wavelength limit, the solutions of this equation must have the same form as (8.33), since such waves will be unaware of the difference between an atomic solid and a continuum. However, the length parameter, which was the continuous variable x in the case of the elastic continuum, must now be the discrete variable ma, where m is an integer, since the displacement is only defined at the atom positions. m is called the running index of the wave. The limiting solution is thus

$$u_m = \xi \exp i(\omega t+mka). \tag{8.36}$$

This trial solution is substituted into (8.35), from which the condition that (8.36) shall be a solution of (8.35) is

$$-\omega^2 M = \beta[\exp ika+\exp(-ika)-2] \tag{8.37}$$

The term in brackets is $4\sin^2(ka/2)$, so the condition (8.37) becomes

$$\omega = \pm\left(\frac{4\beta}{M}\right)^{\frac{1}{2}} \sin\left(\frac{ka}{2}\right) \tag{8.38}$$

where the sign is chosen so that ω is always positive. Equation (8.38) is shown in Fig. 8.2, together with the continuum condition

$$\omega = vk. \tag{8.39}$$

Since the maximum value of $|\sin(ka/2)|$ is unity, there is a maximum frequency that can propagate in the lattice

$$\omega_m = \left(\frac{4\beta}{M}\right)^{\frac{1}{2}}. \tag{8.40}$$

The function (8.38) repeats periodically as k increases.

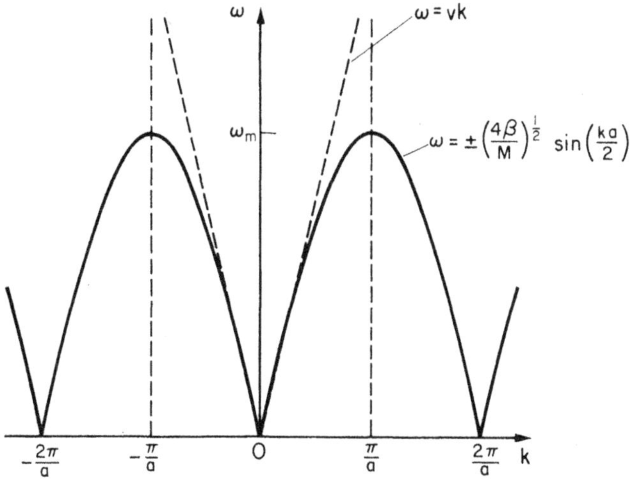

FIG. 8.2. Frequency–wave number relationships in an elastic continuum and in an atomic solid.

The physical significance of this result may be understood by considering the relative motion of two adjacent atoms in the line. For waves of infinite wavelength, their motion is exactly in phase. As the wavelength diminishes, the phase difference between the motions of the two atoms increases, until when the wavelength is twice the distance between them, their motion is exactly out of phase. Every possible relative motion of the two atoms has thus been described in the range of wavelengths $\infty \geqq \lambda \geqq 2a$: further reduction of the length of the waves drawn through the two atoms is possible, but their actual motion is the same as that described by a wave in the range above. Since the motion is only defined at the atom positions, and not in between them, these shorter waves have no physical significance. When it is appreciated that the distance between atoms in a typical solid is only about 3×10^{-8} cm, this restriction on the propagation of elastic waves does not seem very important, but since many of the important processes in solids occur in distances of this sort of magnitude, the restriction is significant.

In terms of the wave number k, the range

$$-\pi/a \leqq k \leqq \pi/a \tag{8.41}$$

contains all the possible different modes of vibration of the lattice, the negative sign corresponding to propagation from right to left, and the positive sign from left to right. This range of values of k is shown by the dotted lines in Fig. 8.2. It should be noted that since (8.38) is periodic in k-space, the "zone" (8.41) is also repeated periodically in k-space. Not only the wavelength is arbitrary, consequently, but also the direction of propagation of a disturbance, since the relative atom motion can be described by a large number of different waves, one in each zone, and some of these will have positive signs and some negative. An example of this is shown in Fig. 8.3. It is usual to remove this ambiguity by restricting the description to a single zone whose centre is defined to be $k = 0$; this restriction has been implied in the discussion above.

A number of workers in the nineteenth century studied this problem of wave propagation, notably Cauchy, Baden-Powell and Kelvin. The modern development of the theory, however, is due largely to Brillouin, and the zone (8.41) in k-space is consequently called the Brillouin zone.

The wave velocity v is now a function of the wave number:

$$v = \omega/k = \frac{1}{k}\sqrt{\left(\frac{4\beta}{M}\right)} \sin \frac{ka}{2} \tag{8.42}$$

In the limit of long wavelength the velocity becomes

$$v_0 = \frac{a}{2}\sqrt{\left(\frac{4\beta}{M}\right)}. \tag{8.43}$$

The line density ρ is now M/a, and the force required to stretch a single bond is

$$F = \beta(u_n - u_{n-1}) = \beta a\varepsilon$$

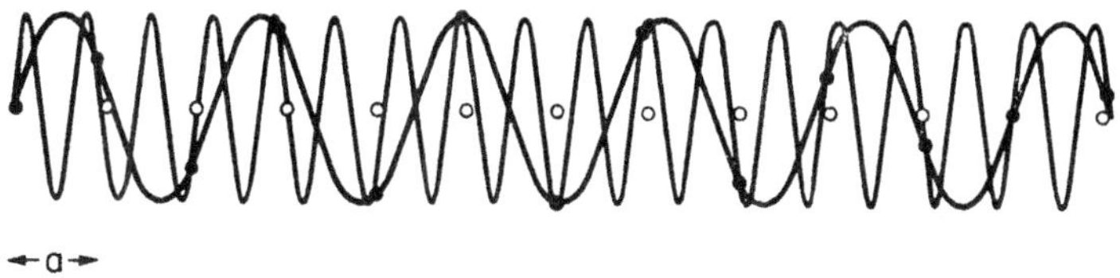

FIG. 8.3. The same atom movements may be described by a number of different waves. Here, the rest positions of the atoms are shown as open circles; their displaced positions by full circles. The longer wave has wavelength $2{\cdot}2a$; so $k = 0{\cdot}91\pi/a$. The shorter wave has wavelength $0{\cdot}69a$; $k = 2{\cdot}91\pi/a$. Thus the two wave numbers differ by exactly $2\pi/a$.

so that recalling that $F = c\varepsilon$, it follows that the force constant is related to the elastic stiffness by the equation

$$c = \beta a. \tag{8.44}$$

Substituting in (8.43) leads to the result that for long waves

$$v_0 = \left(\frac{c}{\rho}\right)^{\frac{1}{2}}$$

which is identical to the result (8.32) for the elastic continuum. In terms of v_0, (8.42) may be written

$$v = v_0\left[\frac{\sin(ka/2)}{(ka/2)}\right]. \tag{8.45}$$

This function is plotted in Fig. 8.4. Consequently, at the Brillouin zone walls, the wave velocity has fallen to the value

$$v_1 = 0{\cdot}6363 v_0. \tag{8.46}$$

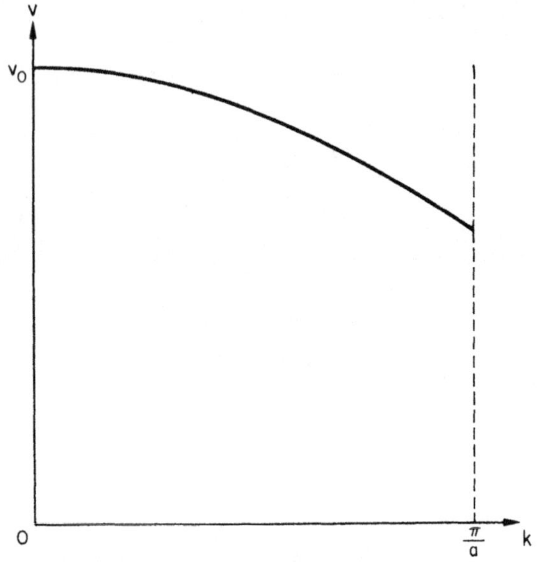

FIG. 8.4. Variation of velocity in the Brillouin zone.

However, we note that at the zone walls the wave function (8.36) has the form

$$u_n = \xi \exp i\omega t \exp in\pi \equiv \xi \exp i\omega t \cos n\pi \tag{8.47}$$

which is the equation for a standing wave, as would be expected from the qualitative argument following (8.40). The apparent anomaly that the wave solution has the form of a standing wave at the Brillouin zone walls, while the wave velocity does not vanish is because the wave solution at the zone walls represents the superimposition of two waves of the same amplitude, but of equal and opposite velocities, each separately equal in magnitude to v_0. This is the situation for total internal reflection of an elastic wave, so it is customary to describe the situation by saying that elastic waves at the Brillouin zone boundaries are totally reflected by the lattice.

The conclusions of the analysis of the linear lattice are thus that waves having frequencies in the range $0 \leqq \omega \leqq \omega_m$ will be passed by the lattice, but that higher frequencies will be attenuated, the amplitude falling off rapidly with distance. In fact, for these higher frequencies the wave number k is imaginary. The physical

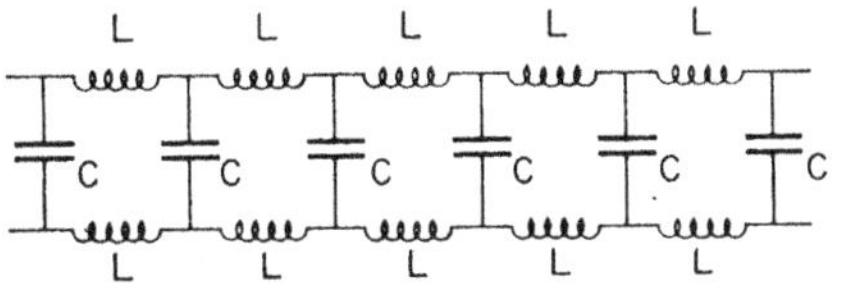

FIG. 8.5. Pupin's low-pass filter.

significance of an imaginary wave number has been discussed in Chapter 5, following (5.38). In electrical circuits the above properties are those of a band-pass filter, and the relevance of the above analysis to the electrical analogue of the loaded cable was first noted by Pupin in 1900, who succeeded in constructing loaded lines and low-pass filters. A typical circuit is shown in Fig. 8.5. The capacitances were in fact provided by the self-capacitance of the cable itself, and the inductances were spaced so that there were about ten inductances per wavelength. The limiting frequency of this arrangement is

$$\nu_m = \frac{1}{\pi\sqrt{(2LC)}}. \tag{8.48}$$

Using exactly the same principle of a periodically loaded line, Campbell in 1906 constructed the first high-pass filter, which passed all frequencies above a certain limit, attenuating those below. A typical circuit is shown in Fig. 8.6. Later, Campbell constructed various band-pass filters, which passed frequencies in a limited range called the passing band, attenuating signals with frequencies higher and lower than the band. A typical circuit is shown in Fig. 8.7. Notice that although the frequency response of a single one of these elements will not itself be linear, some

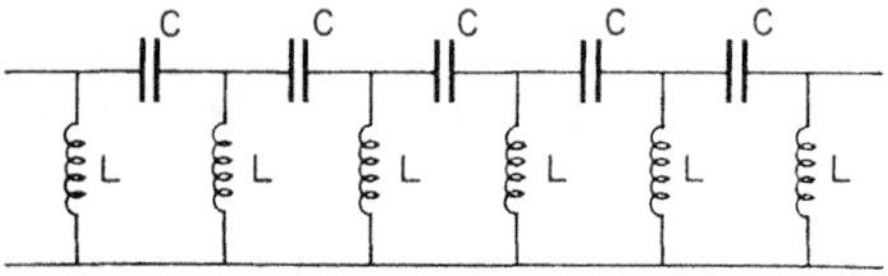

FIG. 8.6. Campbell's high-pass filter.

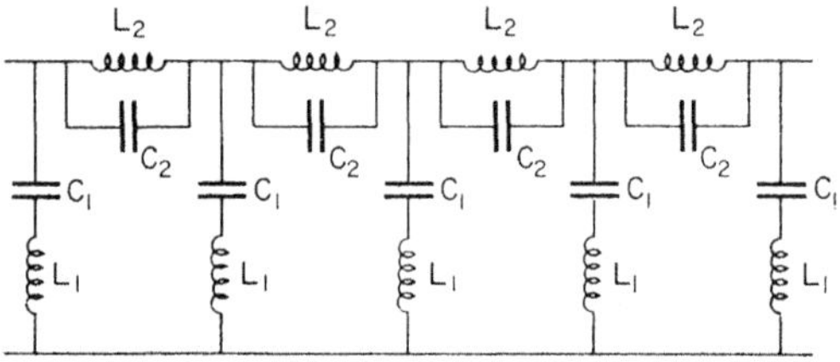

FIG. 8.7. Campbell's band-pass filter.

attenuation taking place at high and low frequencies, the existence of definite band limits is a function of the periodic repetition of the elements. The larger the number of periodic elements in the line, the sharper the definition of the band limits.

A mechanical analogue of band-pass filter was analysed by Kelvin in 1881, but his work went unnoticed. In 1912 Born re-investigated the problem, and quite independently rediscovered Kelvin's analysis. The problem they both considered was the propagation of elastic waves along a linear lattice of two sorts of atom, the linear analogue of a diatomic lattice: Born in fact was expressly interested in the alkali halides. The problem is illustrated in Fig. 8.8. The equations of motion assuming again nearest neighbour interactions are

$$\left.\begin{aligned} m\frac{\partial^2 u_{2n}}{\partial t^2} &= \beta(u_{2n+1}+u_{2n-1}-2u_{2n}) \\ M\frac{\partial^2 u_{2n+1}}{\partial t^2} &= \beta(u_{2n+2}-u_{2n}-2u_{2n+1}) \end{aligned}\right\} \quad (8.49)$$

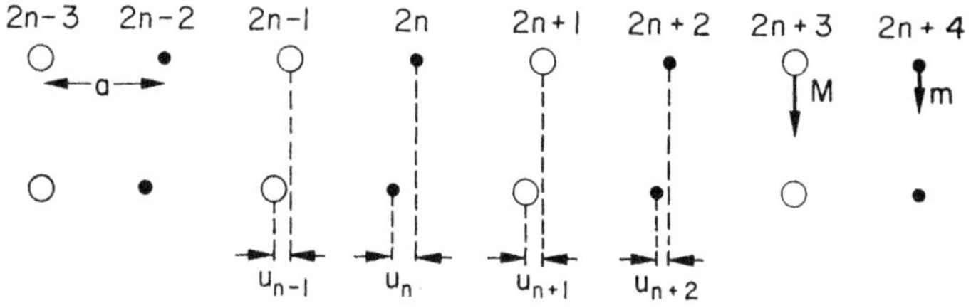

FIG. 8.8. Propagation of an elastic wave along a diatomic line.

where as before u_n represents the displacement of the nth particle. Here, m and M are the masses of the two particles. Again, we look for solutions having the form of (8.36):

$$\left.\begin{aligned} u_{2n} &= \xi \exp i[\omega t+2nka] \\ u_{2n+1} &= \eta \exp i[\omega t+(2n+1)ka]. \end{aligned}\right\} \quad (8.50)$$

These trial solutions are inserted into (8.49), giving the results

$$\left.\begin{aligned} -\omega^2 m\xi &= \beta\eta[\exp ika+\exp(-ika)]-2\beta\xi \\ -\omega^2 M\eta &= \beta\xi[\exp ika+\exp(-ika)]-2\beta\eta. \end{aligned}\right\} \quad (8.51)$$

These two equations are simultaneously satisfied only if

$$\omega^2 = \beta\left(\frac{1}{m}+\frac{1}{M}\right) \pm \beta\left[\left(\frac{1}{m}+\frac{1}{M}\right)^2 - 4\frac{\sin^2 ka}{Mm}\right]^{\frac{1}{2}}. \quad (8.52)$$

For small k (long wavelengths) the roots of (8.52) are

$$\omega^2 = 2\beta\left(\frac{1}{m}+\frac{1}{M}\right) \tag{8.53}$$

and

$$\omega^2 = \frac{2\beta}{M+m}k^2a^2. \tag{8.54}$$

At the Brillouin zone walls $k = \pi/2a$ the roots are

$$\omega^2 = 2\beta/m \tag{8.55}$$

and

$$\omega^2 = 2\beta/M. \tag{8.56}$$

The function (8.52) is plotted in Fig. 8.9 as the full line. Note that if $m = M$, the problem reduces to a single linear chain with spacing a, and hence a Brillouin zone twice the size of that for the diatomic line. The resulting limiting form of (8.52) is the same as (8.38), shown dotted in Fig. 8.8. The effect of altering the masses is to increase the repeat distance along the lattice to $2a$, so the zone is $-\pi/2a \leqq k \leqq \pi/2a$. There are now two passing bands:

$$0 \leqq \omega \leqq (2\beta/M)^{\frac{1}{2}} \quad \text{and} \quad (2\beta/m)^{\frac{1}{2}} \leqq \omega \leqq \left[2\beta\left(\frac{1}{m}+\frac{1}{M}\right)\right]^{\frac{1}{2}}$$

separated by a "stopping band".

The electrical analogue of this problem was studied in 1929 by Mead and French, who were concerned with how to load an overhead cable that was to be joined to an already loaded underground cable. These investigators did not realise that the situation was analogous to the Born problem.

The ratio of the amplitudes of vibration of the two types of atom is ξ/η. For small k, the lower branch gives

$$\eta = \xi \tag{8.57}$$

so that for small k in the lower branch the two types of atom vibrate in phase with the same amplitude, exactly as for the propagation of an elastic wave along a line of similar atoms. For the upper branch at small k,

$$\xi/\eta = -M/m \tag{8.58}$$

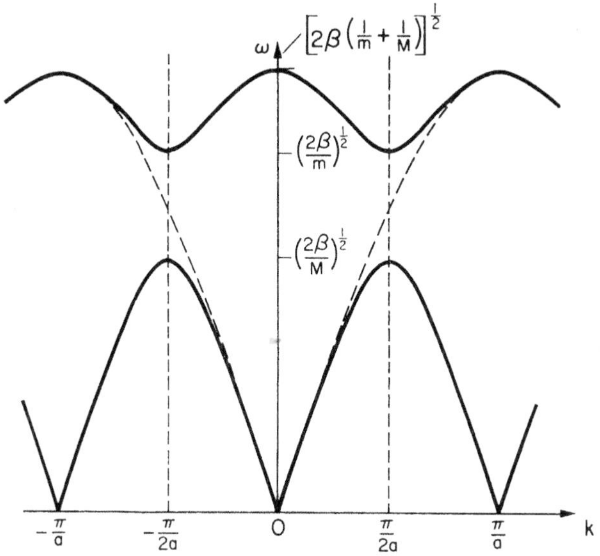

FIG. 8.9. Frequency–wave number relationships in a diatomic solid. (Full line is drawn for $M = 2m$: dashed line shows monatomic limit when $m = M$.)

so that the atoms vibrate out of phase with each other, with amplitudes inversely proportional to their masses. The centre of gravity of a pair of adjacent atoms is consequently at rest during the vibration. However, if the atoms have different charges, as is the case with the alkali halides, the dipole moment of the atom pair oscillates, in contrast with the lower branch, where for small k the dipole moment does not change during the vibration. The oscillating dipole will couple with an electromagnetic wave of the same frequency, so that an electromagnetic wave corresponding to the frequency of the upper branch at $k = 0$ will be strongly absorbed by the crystal. The critical frequency for absorption is thus

$$\omega_0 = \left[2\beta\left(\frac{1}{m}+\frac{1}{M}\right)\right]^{\frac{1}{2}}. \qquad (8.59)$$

For sodium chloride, $\beta \simeq ac_{11}$, where c_{11} is one of the macroscopic elastic stiffness constants of the crystal. $a \simeq 3\times 10^{-8}$ cm, and $c_{11} \simeq 5\times 10^{11}$ dyn cm^{-2}, so $\beta \simeq 1{\cdot}5\times 10^{4}$ dyn cm^{-1}, and

$$\omega_0^2 \cong 2(1{\cdot}5\times 10^4)(1/35{\cdot}5-1/23{\cdot}0)/1{\cdot}67\times 10^{-24}$$

giving

$$\omega_0 \cong 3{\cdot}6\times 10^{13} \text{ rad sec}^{-1}.$$

The wavelength of electromagnetic radiation of this frequency is

$$\lambda_0 = 50\mu \quad (1\mu \equiv 1 \text{ micron} = 10^{-4} \text{ cm}).$$

Experimentally, it turns out that

$$\lambda_0 = 61\mu$$

and in view of the extremely simple model used and the approximations made, this is remarkably good agreement. This wavelength is in the infrared, so the upper band is usually called the optical band. The lower band, because of the absence of an oscillating dipole, does not absorb electromagnetic radiation at $k = 0$, but it does absorb mechanical (acoustical) vibrations: because the centre of gravity of an atom pair remains motionless at $k = 0$ for the upper band this does not absorb acoustical vibrations. The lower band is consequently called the acoustical band. However, at the zone walls the motions described by the two bands are very similar.

In fact, a real atom vibrates in three dimensions, so that the motion of an atom in a linear chain may be described by three waves, two with displacements normal to the length of the chain called the transverse modes, and one with displacements parallel to the length of the chain called the longitudinal mode. The band structure is in general a little different for the transverse and longitudinal modes, although for a linear chain the two transverse modes will be described by the same bands. Qualitatively, however, the arguments are unchanged. In three-dimensional solids the wave equation may be separated into three normal one-dimensional equations, so that again the qualitative statements made above remain valid. Now, however, the resultant wave motion is described by a wave vector **k** which describes not only the wavelength of an elastic disturbance, but also the direction of propagation. As described in Chapter 5, this vector maps out a "**k**-space". Now, the generalisation of the zone limits (8.41) to three dimensions defines a set of planes: the polygonal volume centred on the origin of **k**-space enclosed by these planes is the Brillouin zone for that particular crystal structure. All possible modes of vibration of the three-dimensional lattice are described in terms of wave numbers enclosed in this zone, although as described above the zone may contain a number of bands of permitted frequencies separated by bands of forbidden frequencies. Since the band structure is now a function of direction, it may happen that a frequency which is "forbidden" for one direction of propagation is "permitted" for another direction; in fact, if propagation direction is disregarded, it may happen that no frequency is forbidden. This is described as band overlap.

Finally, it has been assumed hitherto that the line is infinite in length. For a finite line, the situation is altered. Although there will still be passing and stopping bands, which are determined by the periodicity, all the states in a passing band will not be available, because of the finite extent of the line. The structure of the modes depends on the boundary conditions. In the simplest case, a line of length L may be considered clamped at either end; in which case only those waves are permitted for which

$$k = \frac{n\pi}{L} \quad \text{where } n \text{ is an integer.} \tag{8.60}$$

If the interatomic spacing of the line is a, the number of atoms in the line is

$$N = L/a$$

hence, (8.60) becomes

$$k = \frac{n}{N} \cdot \frac{\pi}{a}$$

and since the supplementary band condition is

$$0 \leqq k \leqq \pi/a,$$

it follows that

$$0 \leqq n \leqq N. \qquad (8.61)$$

The case $n = 0$ is trivial, since the line is then at rest; so there are just N possible longitudinal modes of vibration in each band for a linear chain. Including the transverse modes as well, there are altogether $3N$ modes in each band. Since the three-dimensional problem is separable, it follows that this result is true for this case as well.

In conclusion, therefore, the vibration spectrum of a finite three-dimensional atomic solid consists of a band or bands containing $3N$ permitted modes each, where N is the number of identical units (atoms in a monatomic solid, in which case there is only one band; molecules in a molecular solid, in which case there will be more than one band). For a real crystal the number of molecules is very large, so that in fact the levels will be very close together: the band is thus a quasi-continuum. It is important to notice that the band structure is determined only by the periodicity, not by the number of atoms in the assembly.

Although the analysis has been performed for an atomic solid, examples have been quoted of electrical circuit analogues which have exactly the same properties. In fact the derived properties are valid for any situation involving the propagation of a wave in a periodic medium: in such a situation the eigenvalue spectrum consists of quasi-continuous permitted bands separated by forbidden bands: the band limits are functions of propagation direction in the medium and in **k**-space may be represented inside a Brillouin zone which is then periodically repeated.

Quantum Restrictions

So far the problem has been approached in classical terms. However, an atom is a simple harmonic oscillator, for which the energy has one of the values

$$E_{n,k} = (n+\tfrac{1}{2})\hbar\omega_k. \tag{8.62}$$

There are $3N$ values of ω_k for a given solid, associated with the $3N$ values of k, in each band. Each of these modes can have any one of the energies (8.62), and the energy associated with each mode can change by units of $\hbar\omega_k$. By analogy with photons, which are quanta of energy $\hbar\omega$ in the electromagnetic field, the quanta in the elastic field are called phonons. The problem is now to calculate the distribution of the quanta among the various modes ω_k. This problem is dealt with in the next chapter.

The total energy of a crystal is a function of the temperature, and consequently the number and energy of the phonons is a function of the temperature. When heat is conducted through a dielectric solid there are no long range atomic or electronic movements. All that happens is that the atoms vibrate with greater amplitude about their mean positions, the increased amplitude of vibration propagating through the lattice. In the new terminology introduced above, there is a phonon drift through the solid: in fact, for purposes of calculation, there is now no need to consider the atomic nature of the solid; all that need be considered is the properties of the "phonon gas", subject to the limitations of the energy band spectrum. The apparatus for calculating the thermal conductivity of gases is relatively simple, so the introduction of the phonon has already simplified the analysis of solids. The phonon is the first example in this book of a quasi-particle—a concept introduced for convenience, with properties defined in such a way as to simplify the dynamics of the situation. In this case, phonons may be created or destroyed at will, although in this discussion the creation and annihilation of phonons will not be invoked. In addition, unlike ordinary particles, the law of conservation of momentum is not strictly obeyed. For phonons the law of conservation of momentum may be written

$$\mathbf{k}_1 + \mathbf{k}_2 = \mathbf{k}_3. \tag{8.63}$$

But in fact, if $\mathbf{k}_1$ and $\mathbf{k}_2$ are close to the Brillouin walls, (8.63) would carry $\mathbf{k}_3$ outside the zone. In this case, the wave vector must be translated back into the zone by adding or subtracting a vector **G**, where **G** connects equivalent points in two zones. **G** is called a reciprocal lattice vector. The modified law of conservation of momentum is then

$$\mathbf{k}_1 + \mathbf{k}_2 = \mathbf{k}_3 + \mathbf{G} \tag{8.64}$$

so that in fact the collison of two fast phonons may result in the production of one slow phonon, or even one travelling in the opposite direction to the original two. While the original law (8.63) means that phonon–phonon collisions do not affect the average momentum of the gas, the modified collision law (8.64) obviously will produce a net momentum change, and therefore represent a source of thermal resistivity. A collision process which obeys the first law, i.e. one for which **G** is zero, is called a normal process or *N*-process; one for which **G** is not zero is called an umklapp process, or *U*-process. These last were introduced by Peierls in 1929. Apart from *U*-processes, perfectly harmonic phonons should not interact at all, since the displacements of atoms produced by two phonons should superimpose without interaction. However, atom motions, and consequently phonons, are not perfectly harmonic, and the anharmonic terms do lead to phonon–phonon interactions. The analysis of phonon scattering is rather complicated.

Calculation of Thermal Conductivity

Consider the transfer of energy by phonons crossing the *xy*-plane, when a temperature gradient is applied along the *z*-axis. Suppose that the mean free path of the phonons between collisions is Λ. A phonon travelling a distance equal to the mean free path and striking the xy-plane at a polar angle θ has energy

$$E(0)+(\Lambda \cos \theta)\frac{\partial E}{\partial z} \tag{8.65}$$

where $E(0)$ is the mean energy at the plane $z = 0$. The number of phonons crossing unit area of the plane in a direction making an angle between θ and $\theta+d\theta$ with the z-axis per second is

$$\tfrac{1}{2}Nv \cos \theta \, d(\cos \theta) \tag{8.66}$$

where N is the concentration of phonons and v is their average velocity. The factor $\frac{1}{2}$ arises because the random motion of the phonons in thermal equilibrium means that half of the available number will be travelling away from the plane. The net energy flux is then

$$\tfrac{1}{2}Nv \frac{\partial E}{\partial z}\int_{\theta=0}^{\theta=\pi} \Lambda \cos^2 \theta d(\cos \theta) = \tfrac{1}{3} Nv\Lambda \frac{\partial E}{\partial z}. \tag{8.67}$$

But by definition the heat flux is

$$K\left(\frac{\partial T}{\partial z}\right)$$

where K is the thermal conductivity.

Now,

$$N\frac{\partial E}{\partial z} = N\frac{\partial E}{\partial T}\cdot\frac{\partial T}{\partial z} = C\frac{\partial T}{\partial z} \tag{8.68}$$

where C is the heat capacity of the phonon gas per unit volume. Hence, the thermal conductivity is

$$K = (1/3)Cv\Lambda. \tag{8.69}$$

Here, C is the same as the specific heat of the solid, and v is the velocity of the phonons which is approximately the velocity of sound in the medium. There is, of course, a range of phonon velocities, but as demonstrated earlier the slowest phonon travels at 60 per cent of the speed of the fastest, so the range of velocities is small. These quantities may be easily measured, so that the length of the mean free path can be determined. Typical values are 4×10^{-7} cm for quartz at 0°C, rising to 54×10^{-7} cm at -190°C. The corresponding values for sodium chloride are $2{\cdot}3 \times 10^{-7}$ cm and 10×10^{-7} cm. Since a typical lattice spacing in a solid is 3×10^{-8} cm, the mean free paths are relatively very long, of order 100 atomic spacings at low temperatures. This indicates how rare a collision is, because of the weakness of the anharmonic scattering terms.

CHAPTER 9

Statistical Mechanics

IN THE preceding chapters a method has been outlined for determining the possible energies of a dynamical system. However, when the system consists of a large number of components, a complete analysis of the problem is impossible. A complete solution of the dynamics of a system of three components has still to be produced in classical mechanics, and the use of quantum mechanics does not simplify the problem in this respect at all. In a solid there are approximately 10^{23} atoms in every cubic centimetre, so that in fact the mechanics outlined in the earlier section of the book cannot produce exact descriptions of solids. The many-particle problem has to be solved by making a number of simplifying assumptions. The problem consists of two parts: a set of assumptions which enable the possible energy eigenvalues of the system to be evaluated, and a calculation of the distribution of the components of the system among the energy levels. The simplest example of an assumption chosen to make a many-particle system tractable is the so-called "one-particle assumption", in which the energy eigenvalues of the system are assumed to be the same as those of a single particle in the same potential field. It is important to notice that the "one particle" whose behaviour typifies the system may not be the same as a single one of the particles of which the system is composed: it may be necessary to introduce a "quasi-particle" with special properties for the single particle assumption to produce meaningful results. The choice of a potential field for the single-particle model also presents problems, since there will be interactions between the single particle and the remainder of the particles. The model, of course, assumes that a single potential function can be chosen for the system, and since this is plainly impossible if there are strong particle–particle interactions, it follows that the single-particle approximation will not be valid for systems of particles which interact strongly with each other. However, it has already been demonstrated that the interaction between phonons is weak, and of course the long-range interactions between atoms and molecules having inert gas electronic structures are weak, so that there are in fact a number of real situations in which it seems reasonable to use the single-particle approximation. In these cases, it is relatively easy to calculate an energy eigenvalue spectrum which experiment shows is very close to the true energy spectrum. There remains the problem of how the particles in the system are distributed among the energy levels for a given total energy E of the system. This problem requires a new extension of mechanics, called statistical mechanics.

The Basic Problem

Essentially, the problem may be stated in the following terms. Suppose the energy eigenvalue spectrum of the system consists of a series of levels, having energy $e_1, e_2, e_3, \ldots, e_i, \ldots$, etc., and suppose further that n_1 particles have energy e_1, n_2 have energy $e_2, \ldots, n_i$ have energy e_i, etc. The n_i are then termed the population of the ith level. This statement does not presuppose a discrete energy spectrum, since a continuous spectrum may be divided into ranges whose average energies are represented by the e_i. Plainly, the total number of particles in the system is N, where

$$N = \sum_i n_i \tag{9.1}$$

and the total energy of the system E is

$$E = \sum_i n_i e_i \tag{9.2}$$

Now suppose the population of the ith level is altered by a small amount δn_i, with associated small changes in all the other levels, in such a way that both the total energy and the total number of particles in the system remains constant. This stipulation means that

$$\delta N = \sum_i \delta n_i = 0 \tag{9.3}$$

and

$$\delta E = \sum_i e_i \delta n_i = 0. \tag{9.4}$$

Now, in an isolated system of particles of this kind, population changes of this sort will be taking place all the time, as the result of collisions of particles with each other in which energy is interchanged. In such a collision, both particles will move from their original energy levels to new ones, resulting in a change in population of the levels, although, of course, the total energy of the system is unaltered as a result. Some distributions of the particles can be seen to be very unlikely: for example, if the system is arranged so that all the particles have the same energy initially, collisions will raise the energy of some and lower the energy of others: instinctively it can be seen that a return to the original condition as a result of further collisions is unlikely. Similarly, a uniform spreading out of the particles among energy levels around the average will also not be stable. In both cases, there is only one way of arranging the particles for the given distribution. However, in the majority of cases, a given distribution may be achieved by a number of different arrangements of the particles in the energy levels. The system may be regarded as going continually from one arrangement to another, in an entirely random fashion, so that at a given time the distribution of the particles is more likely to correspond to a distribution that can be achieved in many ways rather than one that can only be achieved in one way. The most likely distribution in fact is that for which the number of different arrangements is the largest. In fact, it can be demonstrated that for a system containing a large number of particles this distribution can be achieved in more ways than all the other distributions put together, so that *only* the distribution with the largest number of different arrangements need be determined. This distribution, to which an isolated system will of its own accord change as a result of the internal redistribution from energy transfer processes, is called the equilibrium distribution. The equilibrium distribution is also the most probable distribution, since any arrangement of the particles is equally probable, so that the most probable distribution is that with the largest number of arrangements. There is thus a connection between the thermodynamic concept of equilibrium and the statistical concept of probability. The approach to equilibrium of a system may be equated to a steady increase in probability. It is now necessary to define probability in numerical terms: it is easiest to define the numerical probability of a given distribution as the number of *distinguishable* arrangements corresponding to that distribution. These arrangements are called "complexions". The point about distinguishable arrangements arises because plainly it is possible to "rearrange" the particles in a given energy level; but within the terms of reference such a rearrangement is meaningless. The numerical probability is then

$$W = \frac{N!}{n_1!n_2!\cdots n_i!\cdots} = N!\prod_i (n_i!)^{-1} \tag{9.5}$$

where the top line represents the total number of arrangements of the N particles, and the terms on the lower line represent the indistinguishable rearrangements of the particles within each energy level. The equilibrium distribution is then the one for which W is a maximum:

$$\delta W = 0 \quad \text{for all } \delta n_i. \tag{9.6}$$

Now, probability is a multiplicative property, in the sense that if the probability of a given situation is w_1 and that of another situation is w_2, the combined probability of both situations is $w_1 w_2$. If a quantity S is defined so that

$$S = k_B \ln W \tag{9.7}$$

it is plain that S is an additive property. Here, k_B is some constant.

Combining (9.5) and (9.7),

$$S = k_B(\ln N! - \sum_i \ln n_i!). \tag{9.8}$$

Now, Stirling has shown that if R is large,

$$\ln R! \simeq R \ln R - R \tag{9.9}$$

the error in this approximation being negligible for values of R greater than 100. Using this result, (9.8) becomes

$$S = k_B(N \ln N - N - \sum_i n_i \ln n_i + \sum_i n_i)$$

$$= k_B(N \ln N - \sum_i n_i \ln n_i). \tag{9.10}$$

The condition for equilibrium (9.6) may then be written

$$\frac{\delta S}{k_B} = \sum_i n_i \ln n_i \delta n_i = 0. \tag{9.11}$$

Since the left-hand sides of (9.3), (9.4) and (9.11) are all equal to zero, the equality

$$\sum_i n_i \ln n_i \delta n_i + \lambda \delta N + \eta \delta E = 0 \tag{9.12}$$

holds for all values of the multipliers λ and η, which are consequently undetermined by this equation. Rewriting (9.12):

$$\sum_i (\ln n_i + \lambda + \eta e_i)\delta n_i = 0. \tag{9.13}$$

Since this equality must be true for arbitrary redistributions δn_i, it follows that the term in brackets inside the summation must be equal to zero; and hence

$$n_i = \exp(-\lambda)\exp(-\eta e_i). \tag{9.14}$$

Using (9.1) the total number of particles is

$$N = \exp(-\lambda)\sum_i \exp(-\eta e_i) \tag{9.15}$$

and consequently the *fraction* of the particles on the ith level is

$$\frac{n_i}{N} = \frac{\exp(-\eta e_i)}{\Sigma \exp(-\eta e_i)}. \tag{9.16}$$

In order to evaluate the multiplier η it is necessary to use some other property of the system. The condition of equilibrium for the assembly (9.11) should be compared with the thermodynamic definition of the equilibrium of an isolated system, which is that the "entropy" should be a maximum. Entropy is an additive property, so that it may be conveniently identified with the quantity S that has been introduced in the preceding analysis. This identification is quite general, provided an appropriate choice of the constant k_B is made, because the only two properties of S that have so far been used in the analysis are also shared by entropy. Entropy appears in the thermodynamic definition of temperature:

$$\left(\frac{\partial E}{\partial S}\right)_{N,\ V\ \text{constant}} = T. \tag{9.17}$$

Substituting in (9.12)

$$\eta = \frac{1}{k_B T} \tag{9.18}$$

and so, finally, the law which determines the distribution of the particles of a system among the various available energy levels may be written [from (9.16)]:

$$\frac{n_i}{N} = \exp\left(-\frac{e_i}{k_B T}\right) \Big/ \sum_i \exp\left(-\frac{e_i}{k_B T}\right). \tag{9.19}$$

This distribution law was derived by Maxwell and by Boltzmann, and is called the Maxwell–Boltzmann distribution law: its derivation has been wholly classical, and in particular the Heisenberg Uncertainty Principle has not been invoked. As one might expect, the law breaks down where quantum restrictions become important. The method that has been used to derive the distribution law is called Lagrange's method of undetermined multipliers. Finally, if S is identified with entropy, the quantity k_BT has the dimensions of energy; and it is easy to demonstrate that if N_0 is the number of molecules in a gram-molecule of a perfect gas, then

$$k_B N_0 T = PV \tag{9.20}$$

where P is the pressure of the gas in a volume V. Consequently, the appropriate choice of k_B to identify S with entropy is

$$k_B = R/N_0 \tag{9.21}$$

where R is the gas constant. k_B is called Boltzmann's constant; its value is

$$k_B = 1{\cdot}38047 \times 10^{-16} \text{ erg deg}^{-1}. \tag{9.22}$$

Although the multiplier λ has been eliminated from the distribution law (9.19), it too may be evaluated when S is identified with entropy. The quantity μ, called the "partial molar free energy" or "chemical potential" of a system is defined by the relation

$$\left(\frac{\partial E}{\partial N}\right)_{S,\,V\ \text{constant}} = \mu. \tag{9.23}$$

Substituting this value in (9.12), the remaining undetermined multiplier is evaluated as

$$\lambda = \frac{\mu}{k_B T} \tag{9.24}$$

so that finally the absolute population of an energy level (9.14) may be written in terms of thermodynamic properties of the system:

$$n_i = \exp\left(-\frac{\mu}{k_B T}\right) \exp\left(-\frac{e_i}{k_B T}\right). \tag{9.25}$$

Quantum Statistics

In the preceding section the Maxwell–Boltzmann distribution law has been derived considering only the distribution of the particles among a number of energy levels. However, the energy of a particle in general is composed of a potential energy term and a kinetic energy term, and in a given potential field the state of the particle is wholly determined by specifying six coefficients, three of position; x, y, z; and three of momentum; p_x, p_y and p_z. In classical terms, these coefficients may be specified exactly, but in quantum terms they can only be specified to within a range determined by the experimental accuracy with which they can be determined. Suppose the various errors are δx, δy, δz, δp_x, δp_y and δp_z. It is convenient to imagine the six coordinates as being plotted along six mutually perpendicular axes, the resulting hypothetical space being called a six-dimensional phase space. This device may appear a needless complication, but in reality it simplifies discussion of a number of problems. Now, as the term $\delta x\,\delta y\,\delta z$ may be regarded as a volume element in "volume space" so the term $\delta x\,\delta y\,\delta z\,\delta p_x\,\delta p_y\,\delta p_z$ may be regarded as a volume element in phase space. The state of a particle can be specified only to within this volume. So long as the distribution of points in phase space is diffuse this limitation is not important, having only the same significance as the inevitable experimental error involved in a classical system. Diffuseness in phase space implies either a diffuse distribution in "volume space", that is, low density; or else a low density in momentum space, which in fact implies a high temperature. If the density in phase space becomes high there is a real possibility that two particles may be so close together that the experiment will be unable to differentiate them. In classical terms the problem can be eliminated by performing more accurate experiments, but in terms of quantum mechanics this is not possible. The phase space volume element is in fact the product of three Heisenberg products, so that the minimum volume of the element is of order h^3. Consequently, any system so concentrated in phase space that there is a real possibility that two particles will occupy the same volume element will not be described by the classical distribution law. Such a system is said to be degenerate. A number of different volumes in phase space may correspond to the same energy: consider a system of independent particles in a zero potential field. All points in volume space correspond to the same potential energy, which is in fact zero, and all points on any sphere in momentum space centred on the origin have the same kinetic energy. There are thus a number of "cells" in phase space whose energy is equal. Suppose that there are m_i cells each having energy e_i. Then altering the relative population of the cells will produce a distinguishable arrangement, but since the position of a particle within a cell cannot be specified, redistributions within the cells that leave the populations unaltered cannot be distinguished. The total number of distinguishable arrangements of n_i particles in m_i cells is thus

$$w_i = \frac{(n_i+m_i-1)!}{n_i!(m_i-1)!}. \tag{9.26}$$

The minus one in (9.26) arises because the extreme arrangement corresponds to n_i particles in one cell with (m_i-1) cells empty. The total number of distinguishable arrangements for the whole system are then the products of all the terms like (9.26):

$$W = \prod_i w_i. \tag{9.27}$$

The entropy of the system is then

$$S = k_B \sum_i \ln \frac{(n_i-m_i-1)!}{n_i!(m_i-1)!}. \tag{9.28}$$

Stirling's approximation gives

$$S = k_B \sum_i \left[n_i \ln \frac{n_i+m_i-1}{n_i} + (m_i-1) \ln \frac{n_i+m_i-1}{m_i-1} \right] \tag{9.29}$$

and consequently the equilibrium condition (9.6) becomes

$$\delta S = k_B \sum_i \left\{ \ln \frac{n_i+m_i-1}{n_i} \right\} \delta n_i. \tag{9.30}$$

Again combining this with (9.3) and (9.4), Lagrange's method of undetermined multipliers gives

$$\frac{n_i}{m_i-1} = \frac{1}{\exp(-\mu/k_B T)\exp(e_i/k_B T)-1}. \tag{9.31}$$

If m_i is very large in comparison to 1, as has in fact been implied by the use of Stirling's approximation, the one in the denominator of the left-hand side of (9.31) may be neglected, so that

$$\frac{n_i}{m_i} = \left[\exp\left(-\frac{\mu}{k_B T}\right)\exp\left(\frac{e_i}{k_B T}\right)-1\right]^{-1} \tag{9.32}$$

In order to evaluate this distribution law further, it is necessary to calculate m_i. If the energy levels are close together in comparison to $k_B T$ the energy spectrum may be regarded as a continuum, and the summation sign in (9.28) may be replaced by an integral. In this case, the term n_i/m_i approaches the value dN/dm', where now dm'/dE is the density of states function for the system. For the simple case of a set of independent particles in a box the density of states function has been calculated in (5.59), and for this case

$$dm' = \frac{1}{4\pi^2}\left(\frac{2m}{\hbar^2}\right)^{\frac{3}{2}} E^{\frac{1}{2}} dE. \tag{9.33}$$

Using this value, the distribution law (9.32) becomes

$$N = \frac{1}{4\pi^2}\left(\frac{2m}{\hbar^2}\right)^{\frac{3}{2}} \int_0^\infty \frac{E^{\frac{1}{2}}\, dE}{\exp(-\mu/k_B T)\exp(E/k_B T)-1}. \tag{9.34}$$

Integration of this function gives the distribution of the particles among the energy states as a continuous function. No general integration is possible, but numerical integration may be performed. This distribution law was first derived by Bose to describe the behaviour of light quanta, and by Einstein to describe the behaviour of gas molecules. It is called the Einstein–Bose distribution law. Particles which are described by this law are called "bosons". It should be noted that if

$$N \ll \frac{1}{4\pi^2}\left(\frac{2m}{\hbar^2}\right)^{\frac{3}{2}}$$

the integral in (9.34) must be very small, so that the denominator in the integral must be large. In this case the term unity in the denominator may be disregarded, and consequently

$$N = \frac{1}{4\pi^2}\left(\frac{2m}{\hbar^2}\right)^{\frac{3}{2}} \exp\left(\frac{\mu}{k_B T}\right)\int_0^\infty \exp\left(-\frac{E}{k_B T}\right)E^{\frac{1}{2}}dE \quad (9.35)$$

which is in fact identical with the Maxwell–Boltzmann law for the same situation. The inequality is equivalent to a low density of particles in phase space, so that the Einstein–Bose law is exactly the same as the classical Maxwell–Boltzmann law for low-density distributions in phase space, in agreement with the qualitative analysis above.

The Specific Heat of a Maxwell–Boltzmann Gas

The expression (9.35) may be integrated directly:

$$N = \frac{1}{8\pi^{\frac{3}{2}}}\left(\frac{2m}{\hbar^2}\right)^{\frac{3}{2}} \exp\left(\frac{\mu}{k_B T}\right)(k_B T)^{\frac{3}{2}} \quad (9.36)$$

and the fraction of particles having an energy between E and $E+dE$ is

$$\frac{dN}{N} = \frac{2\exp(-E/k_B T)E^{\frac{1}{2}}dE}{(\sqrt{\pi})(k_B T)^{\frac{3}{2}}} \quad (9.37)$$

The average energy of a particle, E, is plainly

$$\bar{E} = \int_0^\infty \frac{EdN}{N} \quad (9.38)$$

and using (9.37), this becomes

$$\bar{E} = \frac{2}{(\sqrt{\pi})(k_B T)^{\frac{3}{2}}}\int_0^\infty \exp\left(-\frac{E}{k_B T}\right)E^{\frac{3}{2}}dE. \quad (9.39)$$

The integral in this expression is a standard form:

$$\int_0^\infty \exp(-ax)x^n dx = n!/a^{n+1} \tag{9.40}$$

and hence

$$\bar{E} = \tfrac{3}{2}(k_B T) \tag{9.41}$$

Then the specific heat of the gas at constant volume C_v is

$$C_v = \frac{d\bar{E}}{dT} = \tfrac{3}{2}k_B \quad \text{per molecule} \tag{9.42}$$

and multiplying by Avogadro's number N_0 to convert this to practical units;

$$C_v = \tfrac{3}{2}N_0 k_B = \tfrac{3}{2}R \quad \text{per mole.} \tag{9.43}$$

This is in good agreement with the experimental result for monatomic gases. It is interesting to note that the specific heat of a Maxwell–Boltzmann or classical gas is independent of temperature.

The Effect of the Pauli Exclusion Principle

In the quantum statistics outlined above it has been assumed that the particles do not interact, so that the occupancy of a cell in no way alters the probability that another particle will enter it. This is plainly not always so, although in the case of particles in a uniform potential field interactions may well be small. In particular, the interaction implied by the Pauli exclusion principle cannot be ignored. A cell may be identified with a "quantum state", and then the Pauli principle states that each cell on a level may contain only one particle at most. Then if as before there are m_i cells on the ith level, having energy e_i, the number of distinguishable arrangements of n_i particles among the cells is

$$w_i = \frac{m_i!}{n_i!(m_i - n_i)!}. \tag{9.44}$$

Substituting this value in (9.27) and carrying through the calculation as before,

$$\frac{n_i}{m_i} = \frac{1}{\exp(-\mu/k_B T)\exp(e_i/k_B T)+1}. \tag{9.45}$$

This distribution may be applied to a "free gas" of electrons, that is, an assembly of free electrons in which the interactions are assumed to be negligible. However, for this case the density of states function (5.59) must be altered to take account of the double spin degeneracy of an electron:

$$g(E) = \frac{1}{2\pi^2}\left(\frac{2m}{\hbar^2}\right)^{\frac{3}{2}} E^{\frac{1}{2}}. \tag{9.46}$$

For a box of unit volume,

$$m_i = g(E)\delta e_i$$

and so

$$n_i = \frac{1}{2\pi^2}\left(\frac{2m}{\hbar^2}\right)^{\frac{3}{2}} \frac{e_i^{\frac{1}{2}}\delta e_i}{\exp[(e_i-\mu)/k_B T]+1}. \tag{9.47}$$

Summing (9.47) for all the levels,

$$N = \sum_i n_i \cong \frac{1}{2\pi^2}\left(\frac{2m}{\hbar^2}\right)^{\frac{3}{2}} \int_0^\infty \frac{E^{\frac{1}{2}} dE}{\exp[(E-\mu)/k_B T]+1}. \tag{9.48}$$

This distribution law was introduced by Fermi and by Dirac, and is consequently called the Fermi–Dirac distribution law: particles which are described by it (that is, particles which obey the Pauli exclusion principle) are called "fermions". Defining a distribution function $f = n_i/m_i$, the differences between the two quantum mechanical distribution laws may be easily summarised:

$$f_{F-D} = \frac{1}{\exp[(E-\mu)/k_B T]+1} \tag{9.49}$$

and

$$f_{E-B} = \frac{1}{\exp[(E-\mu)/k_B T]-1}. \tag{9.50}$$

Notice that in both cases as the system becomes non-degenerate so that the denominator in the distribution function becomes large, the distribution functions approach the classical value:

$$f_{F-D}, f_{E-B} \to f_{M-B} = \exp\left(\frac{\mu}{k_B T}\right) \exp\left(\frac{-E}{k_B T}\right). \quad (9.51)$$

There are two ways of explaining the approach of both of the quantum statistical laws to the classical law in the limit of low phase space density. In the first place the low density in phase space means that the probability of any cell being occupied is very small, and the probability of any cell being occupied by two or more particles is negligible. In the second place, the difference between particles which obey the Pauli exclusion principle and those which do not is in the eigenvalues of the spin. However, although the translational quantum numbers may approach infinity, the spin quantum number is always restricted to a small set of values. The classical limit may then be regarded as the limit in which Planck's constant is not significantly different from zero: the translational energies and other physical properties associated with the translational degrees of freedom can remain finite as h approaches zero, by letting the appropriate quantum numbers approach infinity. Spin, however, because of the finite limit on the spin quantum number vanishes in the classical limit, and consequently the difference in behaviour between particles having integral spin and those having half integral spin must also vanish.

The Fermi–Dirac Distribution Function

The function f represents the fractional filling of an energy level. When $f = 1$, the level is full, and when $f = 0$, the level is empty. Note that fractional values of f are possible, because there may be many cells, or more properly states, on the same energy level: for fermions, these state are either full or empty, and the value of f gives a measure of the relative proportions which are full and empty. Combined with the density of states function

TABLE 9.1. THE FERMI–DIRAC DISTRIBUTION FUNCTION

$\frac{E-\mu_F}{k_B T}$	$\exp\left\{\frac{E-\mu_F}{k_B T}\right\}$	f	$1-f$	$k_B T f'$
−5	0·0067	0·9933	0·0067	0·0066
−4	0·0183	0·9821	0·0179	0·0177
−3	0·0498	0·9526	0·0474	0·0452
−2	0·1353	0·8808	0·1192	0·1050
−1	0·3679	0·7310	0·2690	0·1966
0	1·0000	0·5000	0·5000	0·2500
1	2·718	0·2690	0·7310	0·1966
2	7·389	0·1192	0·8808	0·1050
3	20·09	0·0474	0·9526	0·0452
4	54·60	0·0179	0·9821	0·0177
5	148·41	0·0067	0·9933	0·0066

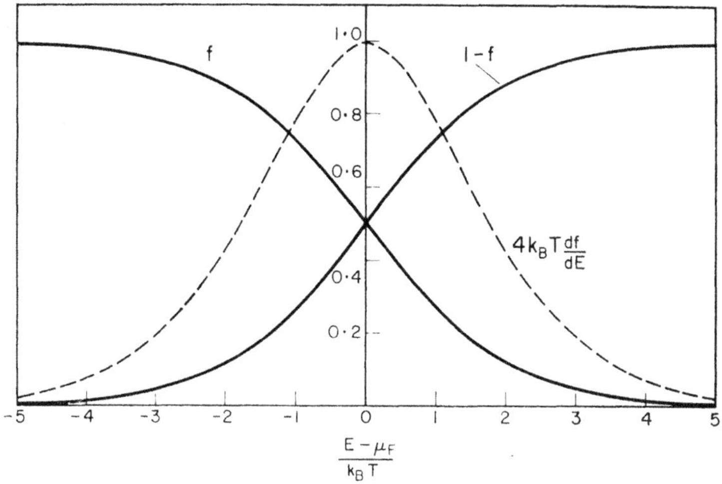

FIG. 9.1. The Fermi–Dirac distribution function.

$g(E)$, the distribution of fermions in the energy spectrum may be directly calculated, using two continuous functions to describe a situation which may in some circumstances be discrete.

When $T = 0$, $f = 1$ for $E < \mu$, and $f = 0$ for $E > \mu$. The chemical potential of the fermions at absolute zero μ_0 is then a cut-off energy, which is called the Fermi energy: to distinguish this from the classical chemical potential the symbol $\mu_{F(0)}$ will be used. At higher temperatures the Fermi energy $\mu_{F(T)}$ is the energy of the level which is exactly half-full, since when $e = \mu_{F(T)}$ $f = 0{\cdot}5$. f approaches unity rapidly for $e < \mu_{F(T)}$, and falls equally rapidly towards zero for values of e greater than the Fermi energy. This is shown in Table 9.1, and it should also be noted that the function $(1-f)$ is a mirror image of the function f. The distribution function is plotted in Fig. 9.1, and the function df/de is also shown. It can be seen that this is only different from zero for a few k_BT about $\mu_{F(T)}$. As with the Einstein–Bose function, no general integration of the Fermi–Dirac distribution law can be given, but numerical integration can be performed, and in some special cases an exact integration may be given.

Integration of the Fermi–Dirac Distribution Function at Absolute Zero

Because $f = 0$ for $E > \mu_{F(0)}$, the upper limit of integration may be altered from ∞ to $\mu_{F(0)}$ without any approximation. Then

$$N = \frac{1}{2\pi^2}\left(\frac{2m}{\hbar^2}\right)^{\frac{3}{2}} \int_0^{\mu_{F(0)}} E^{\frac{1}{2}} dE = \frac{1}{3\pi^2}\left(\frac{2m}{\hbar^2}\right)^{\frac{3}{2}} (\mu_{F(0)})^{\frac{3}{2}}$$

or

$$\mu_{F(0)} = \frac{\hbar^2}{2m}(3\pi^2 N)^{\frac{2}{3}}. \tag{9.52}$$

In numerical terms, this becomes

$$\mu_{F(0)} = 3{\cdot}16 \times 10^{-15}\, N^{2/3}\ \text{eV} \tag{9.53}$$

so that a gas containing 10^{23} fermions per cubic centimetre would have a Fermi energy at absolute zero of 6·79 eV. This means that even at absolute zero some fermions would possess this energy, whereas *all* the particles in a Maxwell–Boltzmann gas have zero energy at absolute zero (9.41). In fact, a Maxwell–Boltzmann gas would have to be at a temperature of $T_{F(0)}$°K where

$$T_{F(0)} = \frac{\mu_{F(0)}}{k_B} \tag{9.54}$$

for the particles to possess this sort of energy. $T_{F(0)}$ is often termed the "Fermi temperature". For the gas of fermions considered above, the Fermi temperature is 78,816°K.

The Temperature Variation of the Fermi Energy at Low Temperatures

By low temperatures in this context is meant

$$T \ll \frac{\mu_F}{k_B}.$$

For dense fermion gases like that considered in the last section this inequality is satisfied for all practical temperatures, but of course for a lower density gas with a correspondingly lower Fermi temperature the inequality may easily cease to be valid at quite low temperatures. The problem of the temperature variation of the Fermi energy in this limit, together with many other problems in the statistics of fermion gases, is simplified by first studying the properties of the integral

$$I = \int_0^\infty f(E) \frac{d}{dE} F(E) dE \tag{9.55}$$

where f is the Fermi–Dirac distribution function and F is any function which vanishes for $E = 0$.

Integrating by parts,

$$I = [f(E)F(E)]_0^\infty - \int_0^\infty F(E)f'(E)dE. \tag{9.56}$$

The term in the square brackets vanishes, because f is zero for $E = \infty$, and F is zero by definition for $E = 0$. Expanding F about μ_F using Taylor's theorem,

$$F(E) = F(\mu_F)+(E-\mu_F)F'(\mu_F)+\tfrac{1}{2}(E-\mu_F)^2F''(\mu_F)+\ldots \quad (9.57)$$

and substituting into (9.56),

$$I = L_0F(\mu_F)+L_1F'(\mu_F)+L_2F''(\mu_F)+\ldots \quad (9.58)$$

where

$$L_0 = -\int_0^\infty f'(E)dE$$

$$L_1 = -\int_0^\infty (E-\mu_F)f'(E)dE$$

$$L_2 = -\tfrac{1}{2}\int_0^\infty (E-\mu_F)^2 f'(E)dE.$$

At low temperatures, the lower limit of integration may be replaced by $-\infty$, since as is shown by Fig. 9.1, the function f' approaches zero rapidly a few k_BT below the Fermi energy, so that so long as the Fermi energy is several k_BT above zero, no appreciable error is introduced by the limit change. Then,

$$L_0 = 1$$

$$L_1 = 0$$

$$L_2 = \tfrac{1}{2}(k_BT)^2\int_{-\infty}^{\infty}\frac{x^2e^xdx}{(1+e^x)^2} = \frac{\pi^2}{6}(k_BT)^2$$

where

$$x = \frac{E-\mu_F}{k_BT}. \quad (9.59)$$

Then (9.55) becomes

$$I = F(\mu_F)+\frac{\pi^2}{6}(k_BT)^2F''(\mu_F)+\cdots \quad (9.60)$$

Now, for the particular problem of the temperature dependence of the Fermi energy, set

$$F(E) = \int_0^E g(E)dE \qquad (9.61)$$

where $g(E)$ is the density of states function, which for the particle in a box problem does fall to zero as E approaches zero. Then

$$N = \int_0^\infty f(E)g(E)dE = \int_0^{\mu_F} g(E)dE + \frac{\pi^2}{6}(k_B T)^2 g'(\mu_F) \,. \qquad (9.62)$$

Subtracting from this the relation

$$N = \int_0^{\mu_{F(0)}} g(E)dE \qquad (9.63)$$

we obtain,

$$\int_{\mu_{F(0)}}^{\mu_F} g(E)dE + \frac{\pi^2}{6}(k_B T)^2 g'(\mu_F) = 0 \qquad (9.64)$$

or, approximately

$$[\mu_F - \mu_{F(0)}]g(\mu_F) + \frac{\pi^2}{6}(k_B T)^2 g'(\mu_F) = 0. \qquad (9.65)$$

Now, substituting (9.46) for $g(E)$,

$$\mu_F = \mu_{F(0)}\left[1 - \frac{\pi^2}{12}\left(\frac{k_B T}{\mu_{F(0)}}\right)^2\right]. \qquad (9.66)$$

This equality will be valid provided the series (9.57) contracts rapidly, and so long as the limit change in (9.58) is permissible. Numerical calculation indicates that in fact (9.66) is approximately valid for $\mu_F > 10k_B T$ at room temperature; this means that fermion gases for which $N > 7{\cdot}4 \times 10^{20}\ \text{cm}^{-3}$ can be considered to have a Fermi energy which is constant and effectively the same as $\mu_{F(0)}$. For lower density gases the temperature dependence of the Fermi energy must be considered.

The Total Energy of a Degenerate Fermion Gas

A fermion gas for which μ_F is very much greater than $k_B T$ for the experimental temperature is called a degenerate fermion gas. For such a gas, the Fermi energy is independent of temperature. The total energy of the gas per unit volume is

$$U = \int_0^\infty E f(E) g(E) dE. \tag{9.67}$$

Now, use the result (9.60) again, this time setting

$$F(E) = \int_0^E E g(E) dE. \tag{9.68}$$

Then

$$U = \int_0^{\mu_{F(0)}} E g(E) dE + \frac{\pi^2}{6}(k_B T)^2 \frac{d}{dE} E g(E). \tag{9.69}$$

Now

$$\int_0^{\mu_{F(0)}} E g(E) dE = U_0,$$

the energy of the gas at $T = 0°$K;

$$= \frac{1}{2\pi^2}\left(\frac{2m}{\hbar^2}\right)^{\frac{3}{2}} \frac{(\mu_{F(0)})^{\frac{5}{2}}}{5/2}. \tag{9.70}$$

But

$$N = \frac{1}{3\pi^2}\left(\frac{2m}{\hbar^2}\right)^{\frac{3}{2}} (\mu_{F(0)})^{\frac{3}{2}} \tag{9.71}$$

and hence

$$U_0 = \frac{3N}{5}\mu_{F(0)}. \tag{9.72}$$

Equation (9.69) becomes

$$U \cong U_0 + (\mu_F - \mu_{F(0)})\mu_{F(0)} g(\mu_{F(0)}) + \frac{3}{2}\frac{\pi^2}{6}(k_B T)^2 g(\mu_{F(0)}). \tag{9.73}$$

Substituting for μ_F,

$$U = U_0 + \frac{\pi^2}{6}(k_B T)^2 g(\mu_{F(0)}). \tag{9.74}$$

The Specific Heat of a Degenerate Fermion Gas

$$C_v = \left(\frac{\partial U}{\partial T}\right)_v = \frac{\pi^2}{3} g(\mu_{F(0)}) k_B^2 T. \tag{9.75}$$

Now, again substituting (9.46) for the density of states function $g(E)$, and (9.53) for N,

$$g(\mu_{F(0)}) = \frac{3N}{2k_B T_F}. \tag{9.76}$$

Finally,

$$C_v = \frac{\pi^2}{2} N k_B \frac{T}{T_F} \quad \text{per unit volume}$$

or

$$C_v = \frac{\pi^2}{2} R \frac{T}{T_F} \quad \text{per mole} \tag{9.77}$$

The specific heat of a fermion gas is thus linearly dependent on temperature (at temperatures low compared to the Fermi temperature) while the Maxwell–Boltzmann gas has a specific heat which is independent of temperature. It should also be noted that since $T \ll T_F$, the specific heat of a fermion gas is very much less than that of a classical gas. The reason for this form of the specific heat is that a gas particle can acquire from thermal sources—that is, as a result of collisions—energies of order $k_B T$. However, only particles within approximately $k_B T$ of the Fermi energy can accept this energy, since otherwise acceptance of the energy takes the particle into a state that is already full. Only near the Fermi energy are there empty states available.

Consequently, only a fraction T/T_F of the fermions can accept thermal energy, so that qualitatively it can be expected that the specific heat of a fermion gas will be approximately T/T_F times the classical value. This is close to the exact result (9.77).

CHAPTER 10

Simple Models of Metals

SOME metals exhibit a very high electrical conductivity, and nearly all metals become very good conductors at very low temperatures. Good conductivity implies a mobile charge carrier, and this has already been briefly discussed in Chapter 7. The charged particles in a metal may be roughly divided into two groups; the positively charged ion cores and the negatively charged valence electrons. Conductivity then implies the motion of one or other of these species. It is difficult to reconcile the observed high conductivity of metals with conduction by ion core drift, and in particular if this were the conduction process, one would expect the conductivity to increase with increasing temperature: in fact the reverse is true for metallic conduction. Conduction by ion drift is observed in ionic crystals, and in this case good agreement is obtained between the mobility of the ions deduced from conductivity measurements, and the mobility deduced from matter-transporting diffusion experiments. In metals, however, the mobility of the ion cores as measured from diffusion experiments is very much less than that required to account for the electrical conductivity. It must be concluded, therefore, that the mobile charged species in metals is the valence electrons. This point was experimentally confirmed by Tolman between 1914 and 1923 who performed an experiment in which a metal specimen was accelerated sharply. As a result of the acceleration a charge difference appeared between opposite ends of the specimen, which Tolman attributed to the inertia of the mobile charge carriers relative to the lattice. Tolman was able to measure the charge and mass of the current carriers, and within the limits of the experiment these agreed with values for the electron.

It, therefore, appears that a metal may be regarded as a lattice of immobile ion cores with valence electrons which exhibit some degree of long-range mobility. On this basis, a model of a metal may be constructed: for simplicity, it is postulated that *all* the valence electrons of the metal are *completely* free, moving as a "free gas" between the ion cores, which are assumed to be spherically symmetrical positive charges at rest on a regular lattice in space. In addition, it will be assumed that the problem can be reduced to the one-electron situation: that is, long-range interactions between electrons are assumed to be negligible. This model is then called the "free electron gas model".

The free electron gas model was first postulated by Drude in 1900, who further assumed that the electron gas was a classical gas whose behaviour could be described by the kinetic theory of gases. The resulting theory is called the "classical free electron gas theory". It is clear that the assumptions in the theory are invalid. If all the valence electrons are in the gas, the gas contains approximately 10^{23} electrons in each cubic centimetre: a classical gas would have to be at a pressure of approximately 10^4 atm to have a density like this, and all gases are exhibiting considerable departures from ideal behaviour at pressures very much less than this. In addition, the average particle to particle distance in the gas is plainly very small, comparable with or less than the inter-atomic distance in the lattice, and since all the electrons are negatively charged, any theory which neglects the existence of strong interparticle Coulomb repulsions would appear to neglect a very large term indeed. Finally, of course, the electrons are anyway not describable in classical terms. The model is thus very much oversimplified, but as will appear, in certain respects it describes quite accurately some properties of real metals.

Some Useful Results of the Kinetic Theory of Gases

Since a Drude metal is described in terms of the kinetic theory of gases, some useful results of the theory will be derived in this section.

Suppose that a gas of non-interacting elastic spheres of radius r moves in a lattice of fixed elastic spheres radius R, where $R > r$. It will be assumed that the mass of the larger spheres is very much greater than that of the smaller spheres, so that as a result of a collision between a small sphere and a large one the larger sphere remains effectively at rest. Plainly, the centres of the two spheres cannot approach closer than a distance $(R+r)$, and if R is much larger than r, this closest distance is not very different from R. The sphere with radius equal to the closest approach distance is called the sphere of influence. If the average velocity of the smaller spheres is $\bar{c}$, then the sphere of influence sweeps out a cylinder of volume $\pi R^2 \bar{c}$ in one second. If there are N large spheres per unit volume this cylinder will contain $NR^2\bar{c}$ large spheres per unit volume, and so there will be this number of collisions in unit time. The average distance the smaller sphere travels between collisions with the larger spheres is thus Λ, where

$$\Lambda = \frac{1}{N\pi R^2}. \qquad \textbf{(10.1)}$$

Λ is called the mean free path. This calculation has neglected collisions between the smaller spheres, since these do not result in a momentum change of the gas as a whole, whereas the collisions of the gas particles with the relatively fixed lattice spheres do constitute a resistance to flow of the gas, altering its average momentum. If it is now assumed that the lattice spheres vibrate with simple harmonic motion about their mean positions, the sphere of influence is the mean area subtended by the vibrating lattice sphere, or

$$R = R_0 + \Delta R \qquad (10.2)$$

where ΔR is half the amplitude of vibration:

$$R = \frac{1}{\pi} a. \qquad (10.3)$$

But

$$a = \frac{1}{\pi \nu}\left(\frac{u}{2m}\right)^{\frac{1}{2}} \qquad (10.4)$$

where u is the energy of the oscillator and ν is the frequency. The vibrational energy of an ion on a lattice site is related to temperature, the mean kinetic energy being $3k_BT/2$. The mean total energy is twice this, so

$$u = 3k_BT. \tag{10.5}$$

Thus

$$\Delta R = \frac{1}{\pi^2\nu}\left(\frac{3k_BT}{2m}\right)^{\frac{1}{2}} = bT^{\frac{1}{2}}. \tag{10.6}$$

The mean free path is consequently a function of temperature:

$$\Lambda = \frac{1}{N\pi R^2} = \frac{1}{N}\left(\frac{1}{R_0^2+2bR_0T^{\frac{1}{2}}+b^2T}\right)$$

$$= \frac{1}{N\pi R_0^2}(1-2bT^{\frac{1}{2}}R_0^{-1}+3b^2TR_0^{-2}+\cdots). \tag{10.7}$$

In a real solid the amplitude of vibration of the atoms at normal temperatures is less than 10 per cent of the interatomic spacing, so

$$bT^{\frac{1}{2}}R_0^{-1} \sim 0.1.$$

To within 1 per cent or so (10.7) reduces to

$$\Lambda = \frac{1}{N\pi R_0^2}(1-2bT^{\frac{1}{2}}R_0^{-1})$$

$$= \Lambda_0(1-2bT^{\frac{1}{2}}R_0^{-1}) \tag{10.8}$$

where Λ_0 is the mean free path at absolute zero. Consequently, according to this model, at room temperature the mean free path will have dropped to about 80 per cent of the value at absolute zero, and will diminish as the square root of the temperature. The rate of change of Λ is relatively slight.

Now set $W(x)$ equal to the probability that the small sphere travels a distance x without a collision. The probability of its travelling a distance $x + dx$ is $W(x + dx)$. But the probability that, having travelled the distance x, the particle will travel the additional distance dx is plainly $W(dx)$; and remembering that probabilities are multiplicative, the probability of the whole journey is $W(x)W(dx)$, which must equal $W(x+dx)$, or

$$W(x+dx) = W(x)+\frac{dW}{dx}dx = W(x)W(dx). \qquad (10.9)$$

Imagine a cylinder length dx, parallel to the direction of the particle, whose cross-section is unity. The probability of a collision within the distance dx is given by the ratio of the combined areas of the front surface through which pass particles which later suffer collisions to the entire front surface area, which by definition is unity. The former quantity is the sum of the projections on the front surface of all the spheres of influence lying within the cylinder, provided dx is small. This is $\pi NR^2\,dx$, so the probability that the section dx is traversed without a collision is given by

$$W(dx) = 1-\pi NR^2\,dx. \qquad (10.10)$$

So, combining (10.9) and (10.10),

$$W(x)+\frac{dW}{dx}dx = W(x)-W(x)\pi NR^2 dx. \qquad (10.11)$$

Integrating,

$$W(x) = C\exp(-\pi NR^2 x). \qquad (10.12)$$

The integration constant C can be evaluated from the fact that the probability is unity for a path of zero length, consequently $C = 1$. Finally,

$$W(x) = \exp(-\pi NR^2 x) = \exp\left(-\frac{x}{\Lambda}\right). \qquad (10.13)$$

This makes it clear that a path between collisions having length exactly equal to Λ is not a particularly frequent event: the distribution of path lengths is exponential, the rate of the exponential decay being characterised by the length Λ.

It is often convenient to define instead a mean free time of flight between collisions τ, given by

$$\bar{c}\tau = \Lambda. \qquad (10.14)$$

The result (10.13) may then be inferred as follows: suppose n_0 particles all suffer a collision at time $t = 0$ and the times of flight of each of them before their next collision is $t_1, t_2, t_3, \ldots$, etc. Then a fraction $n(t)$ will still not have collided again after a time t has elapsed. During the interval of time t to $t+dt$ a fraction $dn(t)/n(t)$ will make their first collision. This is then the probability $W(dt)$, and recalling that

$$\bar{c}dt = dx,$$

and integrating as before,

$$n(t) = n_0 \exp\left(-\frac{t}{\tau}\right). \tag{10.15}$$

In terms of this model the mean free path Λ will be independent of the mean particle velocity in a first approximation, so that

$$\tau \propto (\bar{c})^{-1}. \tag{10.16}$$

Conduction of Electricity by a Drude Metal

In the Drude model, if an electric field $\mathscr{E}$ is applied to the metal, the electron gas experiences an acceleration in the direction of the field. The collision of the electrons with the ion cores acts as a viscous drag on the electrons, and the gas consequently reaches an equilibrium terminal velocity in the direction of the field. This overall "drift velocity" is superimposed on the normal random motion of the electrons, which, of course, has a zero component in any given direction. Suppose the drift velocity is v_D, then

$$v_D = \frac{1}{n} \sum_i v_i \tag{10.17}$$

where the summation extends over all the particles. In the absence of a field, i.e. when $\mathscr{E} = 0$, the motion of the electrons is random, and consequently

$$\sum_i v_{i(0)} = 0. \tag{10.18}$$

Here, $v_i(0)$ is the velocity of the ith electron in the absence of a field. Suppose the field is applied at time $t = 0$. The electron then experiences an acceleration in the direction of the field of $-e\mathscr{E}/m$, so the velocity of the electron at some later time t will be

$$v_i(t) = v_i(0) - \frac{te\mathscr{E}}{m} \tag{10.19}$$

in the absence of a prior collision. The velocity at the next collision will, on average, be

$$v(\tau) = v(0) - \frac{\tau e\mathscr{E}}{m}. \tag{10.20}$$

It is now assumed that the collision randomises the motion of the electron again. It is very difficult to prove this assumption. Consequently, the drift velocity of the electron gas is

$$v_D = \frac{1}{n}\sum_i \left[v_i(0) - \frac{\tau e\mathscr{E}}{m} \right] \tag{10.21}$$

The current density j is then

$$j = nev_D = e\sum_i \left[v_i(0) - \frac{\tau e\mathscr{E}}{m} \right] \tag{10.22}$$

Recalling (10.18), this becomes

$$j = \frac{ne^2\tau\mathscr{E}}{m}. \tag{10.23}$$

This has the form of Ohm's law,

$$j = \sigma\mathscr{E}. \tag{10.24}$$

where the conductivity σ is given by

$$\sigma = \frac{ne^2\tau}{m}. \tag{10.25}$$

However, the mean free time is inversely proportional to the mean velocity $\bar{c}$. Provided the drift velocity due to the applied field is small compared to the random velocities of the electrons due to thermal energies, this implies that the mean free time varies with temperature in the form

$$\tau = \text{constant}\,.\,T^{-\frac{1}{2}}. \tag{10.26a}$$

Experimentally, it is found that the conductivity of a metal decreases linearly with increasing temperature:

$$\sigma = \text{constant}\,.\,T^{-1}. \tag{10.26b}$$

Consequently, the Drude expression (10.25) agrees with experiment only if

$$n = \text{constant}\,.\,T^{-\frac{1}{2}}. \tag{10.27}$$

In other words, the number of free electrons must *decrease* with increasing temperature. This is very hard to understand. In terms of the model, it would be expected that n would be independent of temperature. Even if the assumption that *all* the valence electrons are completely free is abandoned, it would be expected that the number of free electrons would *increase* with increasing temperature. Consequently, although Drude's theory does predict an ohmic conductivity for metals, the experimental temperature dependence of the conductivity is incompatible with the theory.

Numerically, for copper which has one valence electron per atom,

$$n = \frac{\text{Avogadro's number}}{\text{gram atomic volume}} = \frac{6{\cdot}025\times 10^{23}\times 8{\cdot}94}{63{\cdot}5}$$

$$= 8{\cdot}5\times 10^{22}\ \text{cm}^{-3}$$

$$\Lambda = 1/N\pi R^2 \cong 4\times 10^{-8}\ \text{cm}$$

$$\tau = \Lambda/\bar{c} = 3{\cdot}47\times 10^{-15}\ \text{sec.}$$

For calculating these last values, R is known from scattering experiments to be about $0{\cdot}96\times10^{-8}$ cm for copper, $N = n$, and the mean velocity $\bar{c}$ is given by kinetic theory as

$$\bar{c} = \left(\frac{3k_BT}{m}\right)^{\frac{1}{2}} = \left(\frac{3\times1{\cdot}38\times10^{-16}\times300}{9{\cdot}109\times10^{-28}}\right)^{\frac{1}{2}}$$

$$= 1{\cdot}17\times10^{7}\ \text{cm sec}^{-1}.$$

Finally therefore

$$\sigma = \frac{ne^2\tau}{m} = \frac{e^2}{m\bar{c}R^2\pi}$$

$$= \frac{23{\cdot}1\times10^{-20}}{9{\cdot}11\times10^{-28}\times1{\cdot}17\times10^{7}\times3{\cdot}142\times0{\cdot}922\times10^{-16}}$$

$$= 8{\cdot}3\times10^{4}\ (\text{ohm cm})^{-1}.$$

This compares with an experimental value of about 6×10^{5} $(\text{ohm cm})^{-1}$.

Suppose that the electric field has been applied to the metal for a very long time, so that the assembly has reached a steady state, with a uniform drift velocity of the electron gas $v_D(0)$. Now let the field be abruptly reduced to zero at time $t = 0$. Collisions will then slow the gas down, gradually reducing the drift velocity and returning the velocity distribution in the gas to the random value. If again it is assumed that one collision is sufficient to randomise the motion of an electron, the drift velocity $v_D(t)$ at some later time t is

$$v_D(t) = v_D(0)\exp\left(-\frac{t}{\tau}\right) \qquad (10.28)$$

since from (10.15) a fraction of the electrons $\exp(-t/\tau)$ have lost the field induced drift due to collisions, the remaining part of the gas retaining the original drift velocity. This type of exponential decay of an initial disturbance is well known, and is called "relaxation". It will be noticed that in (10.28) τ appears as a scaling term, in effect defining the unit of time for the relaxation process. For this reason, it is often called the "characteristic time" of the process, or alternatively the "relaxation time".

The Conduction of Heat by a Drude Metal

From the observed fact that generally speaking metals that are good conductors of electricity are also good conductors of heat, Drude postulated that heat was also transported through the metal by the electron gas. Now, unlike the electron in an electric field, an electron in a thermal gradient does *not* continuously accelerate. It can alter its energy, and hence its velocity, only as a result of a collision with a lattice ion core. It is the energy of these which alters as a result of the applied thermal gradient.

Consider a metal in which there is an applied thermal gradient $\partial T/\partial x$. The average velocity $\bar{c}$ and energy E of the electron are now functions of the coordinate x:

$$\bar{c} = \bar{c}(x)$$

and

$$E = E(x).$$

An electron whose direction of motion lies at an angle θ to the x-axis collides with an ion lying somewhere within a unit area in the plane $x = x_0$. It will have made its previous collision with an ion lying in the plane

$$x_1 = x_0 - 2\tau\bar{c}\cos\theta \tag{10.29}$$

and consequently will have an energy immediately prior to the second collision of $E(x_1)$.

The number of free electrons per unit volume whose directions of flight lie in a solid angle $d\omega$ is $n d\omega/4\pi$, so that the number whose directions of flight make angles with the x-axis lying between θ and $\theta + d\theta$ is

$$\tfrac{1}{2}n \sin\theta \, d\theta.$$

The number of free electrons satisfying this condition and passing through unit area in the plane x_0 in time dt is equal to the number contained in a cylinder with unit base in the plane x_0 and a slant height $\bar{c}dt$. The perpendicular height will thus be $\bar{c}\cos\theta\, dt$, and since the base has unit area, this is also the volume of the cylinder. Finally, then, the energy flux across unit area in the plane $x = x_0$ in time dt is

$$dt \int_0^\pi E(x_1)\tfrac{1}{2}n\bar{c}\cos\theta\sin\theta\, d\theta$$

$$= dt \int_0^\pi \left[E(x_0) - 2\tau\bar{c}\cos\theta\left(\frac{\partial E}{\partial x}\right)_{x=x_0}\right]\tfrac{1}{2}n\bar{c}\cos\theta\sin\theta\, d\theta$$

$$= -\tfrac{2}{3}dt\, n\tau\bar{c}^2\left(\frac{\partial E}{\partial x}\right)_{x=x_0}. \tag{10.30}$$

In a metal of uniform composition, E can depend on x only through the temperature T, so that

$$\frac{\partial E}{\partial x} = \left(\frac{dE}{dT}\right)\frac{\partial T}{\partial x}. \tag{10.31}$$

The thermal current density is thus

$$w = -\tfrac{2}{3}n\tau\bar{c}^2\frac{dE}{dT}\frac{\partial T}{\partial x} \tag{10.32}$$

so the thermal conductivity, defined as

$$K = -w\bigg/\frac{\partial T}{\partial x} \tag{10.33}$$

is given by the Drude theory as

$$K = \tfrac{2}{3}n\tau\bar{c}^2\frac{dE}{dT}. \tag{10.34}$$

But

$$n\frac{dE}{dT} = C_v, \tag{10.35}$$

the specific heat of the electron gas at constant volume. Now, with Drude, the specific heat of the electron gas is calculated in terms of the kinetic theory of gases:

$$E = \tfrac{1}{2}m\bar{c}^2 = \tfrac{3}{2}k_B T \tag{10.36}$$

$$C_v = \tfrac{3}{2}k_B n \tag{10.37}$$

so,

$$K = \frac{3k_B^2 n\tau}{m} T. \tag{10.38}$$

TABLE 10.1. EXPERIMENTAL VALUES OF THE LORENTZ NUMBER

Group	Metal	L (W Ω deg^{-2} × 10^8) 0°C	100°C
IB	Cu	2·23	2·33
	Ag	2·31	2·37
	Au	2·35	2·40
IIB	Zn	2·31	2·33
	Cd	2·42	2·43
IVB	Pb	2·47	2·56
	Sn	2·52	2·49
VIA	Mo	2·61	2·79
	W	3·04	3·20

This produces an interesting result. The ratio of the electrical conductivity (10.25) and the thermal conductivity (10.38) of a Drude metal are

$$\frac{K}{\sigma} = 3\left(\frac{k_B}{e}\right)^2 T. \tag{10.39}$$

Now in fact Wiedemann and Franz in 1853 had noted that at

room temperature the ratio of the thermal and electrical conductivities of most metals was approximately the same, and in 1872 Lorentz had discovered that this ratio was a linear function of the absolute temperature. Empirically, therefore, most metals obeyed the Wiedemann–Franz law:

$$\frac{K}{\sigma} = LT \tag{10.40}$$

where L, the Lorentz number, is tabulated for some metals in Table 10.1. It can be seen that (10.40) is only approximately true, and that in particular L appears to show some dependence on position in the Periodic Table. According to the Drude theory

$$L_{\text{th}} = 3\left(\frac{k_B}{e}\right)^2 = 2{\cdot}23\times 10^{-8}\ \text{W}\ \Omega\ \text{deg}^{-2} \tag{10.41}$$

This is very close indeed to the experimental values.

The Specific Heat of a Drude Metal

In a Drude metal there are two components: the lattice of ion cores, which vibrate with simple harmonic motion about their mean positions, and a gas of free electrons, which behaves as a classical gas. Consequently, the specific heat should contain two terms, one representing the specific heat of the lattice and one the specific heat of the electron gas:

$$(C_v)_{\text{total}} = (C_v)_{\text{lattice}} + (C_v)_{\text{electron gas}}. \tag{10.42}$$

The lattice specific heat has already been calculated: it is

$$(C_v)_{\text{lattice}} = 3Nk_B \tag{10.43}$$

and the specific heat of a classical gas is well known from the kinetic theory:

$$(C_v)_{\text{gas}} = \tfrac{3}{2}Nzk_B. \tag{10.44}$$

Here, z is the number of free electrons per atom. According to Drude, the specific heat of a metal should be

$$(C_v)_{\text{total}} = 3Nk_B\left(1+\frac{z}{2}\right) \text{ per unit volume, or}$$

$$3R\left(1+\frac{z}{2}\right) \text{ per mole,} \qquad (10.45)$$

whereas an insulator, with no free electrons, should have a specific heat of only $3R$ per mole. In fact, as was pointed out by Dulong and Petit, most monatomic solids have specific heats close to $3R$ per mole at high temperatures, and there is no large difference between metals and insulators. It appears therefore that the specific heat of the electron gas is very much smaller than that of a classical gas. Careful measurement shows that there is a small term in the specific heat of metals that is absent from that of insulators, but it is about 100 times less than the second term in (10.45) and is linearly dependent on the absolute temperature.

Lorentz's Modification of the Drude Model

Although the Drude model explains the Wiedemann–Franz law, and predicts a value of the Lorentz number which is very close to the experimental value, the temperature dependence of the conductivity is altogether wrong, and the numerical value of the conductivity predicted by the theory is in poor agreement with experiment. In particular, the theory is incapable of accounting for the very high conductivities of most metals close to absolute zero. Lorentz modified the Drude theory, making a much more careful calculation of the collision processes using a method due to Boltzmann, and using Maxwell–Boltzmann statistics to describe the electron gas rather than the kinetic theory of gases. The effect of this improvement in the mathematical apparatus of the theory was to demonstrate explicitly that the model was incapable of giving the correct temperature dependence of the electrical conductivity, and diminishing the numerical agreement between the theoretical and experimental Lorentz numbers. The specific heat anomaly was unaltered.

Sommerfeld's Model of a Metal

The potential experienced by an electron in a metal is composed of two parts: that due to the lattice of positively charged ion cores, and that due to all the other electrons. Sommerfeld assumed that the effect of the electrons was to smooth out the periodic potential due to the ion cores, so that the resultant potential is uniform throughout the metal, changing abruptly at the boundaries. This potential is simply that of the "three-dimensional potential box" discussed in Chapter 5. Clearly also, by assuming that the effect on a given electron of all the other electrons can be represented by a single time independent function, the effect of specific electron–electron interactions are excluded, so that the Sommerfeld model is a "one-electron" model. Sommerfeld then applied wave mechanics to the single electron in the potential box, deriving an eigenvalue energy spectrum which is then assumed to be the same for all the electrons. Then, because of the restrictions imposed by the Pauli exclusion principle each energy level can accommodate one electron at most: at absolute zero it is reasonable to assume that the electrons are in the lowest available energy states, so that if there are N free electrons in the metal, the N lowest energy states, calculated for a single electron, will be filled: all higher energy states will be empty. In effect, therefore, Sommerfeld has used a one-electron method to obtain an energy eigenvalue spectrum, and then describes the electron gas using Fermi–Dirac statistics instead of the kinetic theory used by Drude and the Maxwell–Boltzmann statistics used by Lorentz.

The energy levels are

$$E = \frac{\hbar^2 \pi^2}{2ma^2} |\mathbf{n}|^2 \tag{5.55}$$

where a is the side of the box—in this case, the side of the piece of metal—and **n** is the quantum vector. The surfaces of constant energy form a set of concentric spheres in **n**- or **k**-space. Because of the Pauli exclusion principle each point defined by three integers n_x, n_y and n_z can accommodate only two electrons, one of each spin. Consequently, at absolute zero all the points within a sphere radius $|\mathbf{k}_F|$ are occupied, all states outside this sphere are empty. Plainly, the energy of this spherical surface is the Fermi energy μ_F, and the spherical surface is called the Fermi surface. In a metal with approximately 10^{23} electrons per cubic centimetre, the Fermi energy is in the range 5–10 eV. As the temperature is increased electrons can increase their energy as a result of collisions with excited lattice atoms. However, an electron can only gain amounts of energy of order k_BT in this way, and can only accept it if a state corresponding to its new energy is vacant. Consequently, only states within a few k_BT of the Fermi surface can accept energy from the lattice. Since k_BT at room temperature is only about 0·025 eV, very few of the electrons can accept energy, and the properties of the electron gas are thus largely determined by the fraction of electrons in the immediate vicinity of the Fermi surface.

One of the consequences of this is that only a fraction

$$\frac{2k_BT}{\mu_F}$$

can contribute to the specific heat. The electronic contribution to the specific heat will thus be approximately

$$(C_V)_{\text{electron gas}} = \frac{3Rk_B}{\mu_F}T. \tag{10.46}$$

This is approximately 200 times less than the classical value, and is linearly dependent on temperature, in good agreement with experiment. Sommerfeld's model thus removes one of the major anomalies of the Drude theory, but cannot really be regarded as removing the difficulties connected with the temperature dependence of the resistivity, since the collision mechanism is not expressly dealt with in Sommerfeld's theory.

Quantitative Treatment of the Sommerfeld Model

The qualitative argument presented above can be formalised in terms of the equations we have already derived. The density of states function for the Sommerfeld model is

$$g(E) = \frac{1}{2\pi^2}\left(\frac{2m}{\hbar^2}\right)^{\frac{3}{2}} E^{\frac{1}{2}} \tag{9.46}$$

and the states are filled according to the Fermi–Dirac distribution law:

$$f(E) = \frac{1}{\exp\left[(E-\mu_F)/k_B T\right]+1}\,. \tag{9.49}$$

Since $\mu_F \gg k_B T$ for most normal temperatures the distribution is highly degenerate. The temperature dependence of the Fermi energy is thus very small (9.66), and so $\mu_F \sim \mu_{F(0)}$. The specific heat of the fermion gas is

$$(C_v)_{\text{fermion gas}} = \frac{\pi^2}{2}\frac{Rk_B}{\mu_{F(0)}}T \tag{9.77}$$

in close agreement with the result of the qualitative analysis above.

In any process in which an electron collides its momentum is changed, and as a result only collisions are permitted in which an electron initially on the Fermi surface is transferred to another position on the Fermi surface. All electrons involved in collisions consequently have the same speed, called the Fermi velocity:

$$v_F = \sqrt{\left(\frac{2\mu_F}{m}\right)}\,. \tag{10.47}$$

For a metal with $\mu_F = 5$ eV, this is approximately $1{\cdot}35\times10^8$ cm sec^{-1}. The fact that the electrons involved in the conduction process all have the same velocity explains why the Drude theory, which uses the single velocity of the kinetic theory, gives better numerical agreement than the Lorentz theory, which uses the range of velocities of the Maxwell–Boltzmann distribution. The effect of a field is to accelerate the electron gas, reaching a terminal drift velocity v_D parallel to the field, the acceleration being opposed by the scattering processes at the Fermi surface. In terms of **k**-space, this means that the entire Fermi sphere is displaced from the origin. The drift velocity of the electron gas is given by the position of the centre of the sphere, so that if the field $\mathscr{E}$ is applied in the x-direction, the centre of the sphere is displaced from the origin to a position $(k_x, 0, 0)$ where

$$k_x = \frac{mv_D}{\hbar}. \tag{10.48}$$

The mobility of the electron gas η is the drift velocity for unit field:

$$\eta = \frac{v_D}{\mathscr{E}} \tag{10.49}$$

and for copper this is approximately 30 $cm^2 \, V^{-1} \, sec^{-1}$, so that for a field of 1 V cm^{-1} the drift velocity is 30 cm sec^{-1}. Equation (10.48) is then approximately 27 cm^{-1}, compared to $k_F \sim 10^5$ cm^{-1}. Consequently, the displacement of the Fermi sphere for all normal fields is very small.

Relaxation in the Sommerfeld Model

Sommerfeld chooses to regard relaxation as an empirical process, characterised by a relaxation time τ. He does not explicitly consider the nature of the collision process. Using the result

$$\tau = \frac{m\sigma}{ne^2} \tag{10.25}$$

for copper,

$$\sigma = 6{\cdot}5 \times 10^5 \, (\Omega \, \text{cm})^{-1} \text{ at room temperature}$$

$$n = 8{\cdot}5 \times 10^{22} \, \text{cm}^{-3}$$

and hence

$$\tau \simeq 2 \times 10^{-14} \text{ sec.}$$

Remembering that the velocity of the electrons is about $1{\cdot}6\times10^8$ cm sec^{-1}, the mean free path becomes

$$\Lambda \cong 3{\cdot}2\times10^{-6}\ \text{cm} = 320\ \text{Å}.$$

There is good reason to believe that this large value is correct. Reynolds and Stilwell in 1952 measured the resistivity in thin films. By observing when scattering at the surface of the sample began to dominate the ordinary scattering in the bulk of the material as a function of film thickness, they were able to estimate the length of the mean free path in the bulk metal, obtaining a value of 450 Å for copper at room temperature. The mean free path increases as the temperature is lowered, approaching infinity at absolute zero. While this is not a defect in Sommerfeld's theory, it is unfortunate that the model gives no explanation for the scattering process.

Detailed calculation of the thermal conductivity of a Fermi–Dirac gas with a relaxation time at the Fermi surface of τ gives the result:

$$K = \frac{\pi^2}{3}\,k_B^2 T\,\frac{n\tau}{m} \qquad (10.50)$$

and consequently the Lorentz number L becomes

$$L = \frac{\pi^2}{3}\left(\frac{k_B}{e}\right)^2 = 2{\cdot}45\times10^{-8}\ \text{W}\ \Omega\ \text{deg}^{-2} \qquad (10.51)$$

in even closer agreement with experiment than (10.41), although both lie well within the scatter band of experimental results.

It follows, therefore, that the Sommerfeld free electron model produces a satisfactory agreement with experiment; but fails to account explicitly for the relaxation process. Consequently, the reason for the temperature dependence of the electrical conductivity of metals is still not clear. In addition, the model suggests no reason why some materials are metallic conductors, some semiconductors, and some insulators; and it does not explain the fact that although group II metals such as zinc have *two* valence electrons per atom, they are nevertheless considerably worse conductors than group I metals, such as copper, which only have one valence electron per atom.

CHAPTER 11

The Band Theory of Solids

IN CHAPTER 7 the energy levels of electrons in multi-atom systems were briefly considered, and it appeared that so long as the "one-electron approximation" remains valid, each single level of the electron in a single atom splits into N levels in a system of N atoms. It also appears, from the functions sketched in Figs.7.6 and 7.9, that the magnitude of the splitting depends on the closeness of approach of the atoms, so that in fact the remainder of the N levels lie between the two extreme levels whose position is entirely determined by the closest distance of approach between two atoms in the assembly, the "nearest-neighbour" distance. In addition, the analysis of Chapter 7 makes it clear that the magnitude of the splitting is a function of the overlap between the wave functions corresponding to the state in the single atom: consequently, for a given nearest-neighbour spacing, the magnitude of the splitting is greater for states with functions with greater radial extent.

These qualitative arguments can be formalised in terms of a diagram showing the variation of the energy eigenvalues of a single electron in an assembly of atoms as a function of the nearest-neighbour distance a. This diagram is shown in Fig. 11.1. The shaded portions show the splitting of the levels. Each shaded portion represents $N \times \delta$ levels, where δ is the number of states with the same energy level in the free atom—the degeneracy of the level. For an s state, $\delta = 2$; for p states, $\delta = 6$; for d states, $\delta = 10$; etc. In a real specimen, N will be of the order of 10^{23}. The result is that the levels must be very closely packed. It is inconceivable that the total width of any split level at equilibrium nearest-neighbour distance could be as great as 100 eV, and in fact a more typical figure would be 10 eV: however, to take an extreme case, it is known from the sharpness of the characteristic X-ray spectral emission lines from solids that the innermost levels are virtually unsplit, and that they lie approximately 10^4 eV below the uppermost levels. Consequently, the

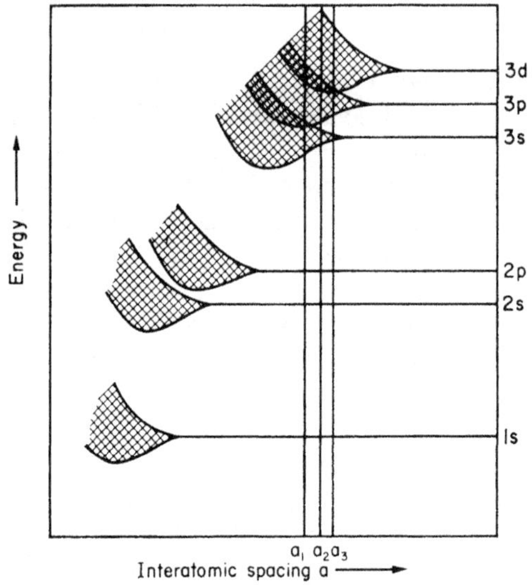

FIG. 11.1. The splitting of energy levels in the solid.

absolute maximum width of the split levels cannot be greater than 10^4 eV. Since in this range there are approximately 10^{23} levels, it follows that the average energy difference between adjacent levels is less than 10^{-19} eV. This is a very small energy difference. In order for an electron to have difficulty acquiring this amount of energy from thermal sources, the temperature would have to be reduced to less than 10^{-17}°K. It follows then that under all real circumstances an electron will be able to move freely from one energy level to another (provided, of course, the level to which it moves is not already occupied). This is the property of the classical "energy continuum" in which the energy of a particle may be continuously varied, so that within the shaded regions in Fig. 11.1 the energy is effectively continuous. Since, unlike the continuum, the number of states in a finite energy range is finite, this situation is termed a "quasi-continuum". The split state is referred to as an energy band, and the width E_B that the band has for a nearest-neighbour distance a_0 is termed the band-width.

In fact, the band-width is non-zero for all states for all finite atom spacings: but in practice it is very narrow until the overlap between the wave functions becomes appreciable. The band width, is should be remembered, is a function only of the degree of overlap between the wave functions, and hence of the nearest-neighbour distance: it does not depend at all on the number of atoms in the assembly. This means that many important properties of the band spectrum and hence of the solid itself will be independent of specimen size.

Variation of Band Spectrum with Atomic Spacing

In practice, the largest electronic contribution to the binding energy of a solid comes from the outermost filled or partly filled states, which are those states arising from the valence states of the free atoms. Consider, for example, with the aid of Fig. 11.1, a solid composed of atoms having valence electrons in 3*s* and 3*p* states. The equilibrium spacing of the atoms in the solid will have a value close to a_1: it can be seen that the inner levels form bands whose width is negligible, with energies very close to the free atom values. This is in agreement with experiment, since not only is the characteristic X-ray emission spectrum, which arises as a result of excitation from these inner levels, sharp; it is also independent of the state of combination of the atoms, being the same for a free atom, a liquid, a solid or a chemical compound of the atom. However, the levels associated with the valence electrons are considerably split into bands. Again, there is direct experimental evidence that this is so: the

X-ray emission spectrum arising from the excitation of electrons in these states consists of long-wave radiation, the so-called "soft" X-rays. The soft X-ray spectrum is dependent on the state of aggregation of the atoms, and for a solid it consists of broad

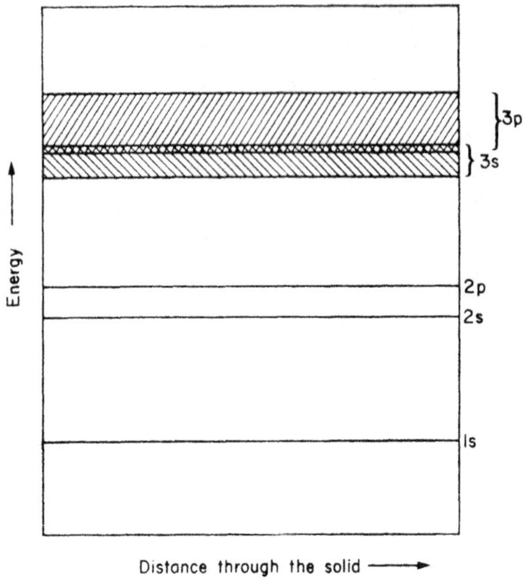

FIG. 11.2. Energy spectrum for a solid in which the interatomic spacing is a_1.

diffuse lines, entirely compatible with the concept of a band spectrum.

The energy spectrum in the solid for three typical nearest-neighbour distances a_1, a_2 and a_3 are sketched in Figs. 11.2, 3 and 4, using the energies shown in Fig. 11.1. In these diagrams the energies of the various bands are shown against a parameter labelled "distance through the solid". In terms of the model, the energy spectrum will be uniform along a given direction in a homogeneous solid, so that the spectrum is shown as uniform horizontal bands. However, this method of plotting is useful for two reasons: firstly, because any local inhomogeneity may be represented as a local distortion of the band structure, and an applied field may be represented by tilting the bands relative to the distance axis; and secondly because it helps to remind one that

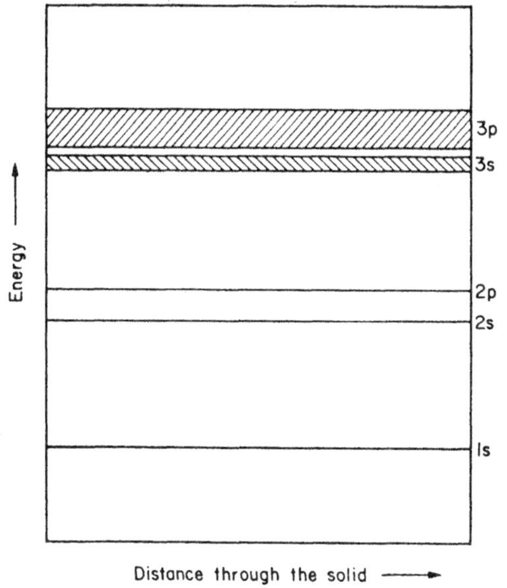

FIG. 11.3. Energy spectrum for a solid in which the interatomic spacing is a_2.

the band structure is different for each direction through the crystal, since the degree of overlap will also be different for different directions.

Electrical Properties of Solids in Terms of Band Theory

The diagrams so constructed are called band diagrams. Customarily, only the top two bands are shown, the part of the spectrum which closely resembles the free atom spectrum being omitted. In the first case, Fig. 11.2, the bands derived from the 3*s* and 3*p* states overlap, forming a composite quasi-continuous band having $8N$ states. At the bottom of the band the states are pure 3*s* states; at the top of the band they are pure 3*p* states; but between it cannot be said that any particular state within the band is a 3*s* or a 3*p* state. Such states are called hybrid states,

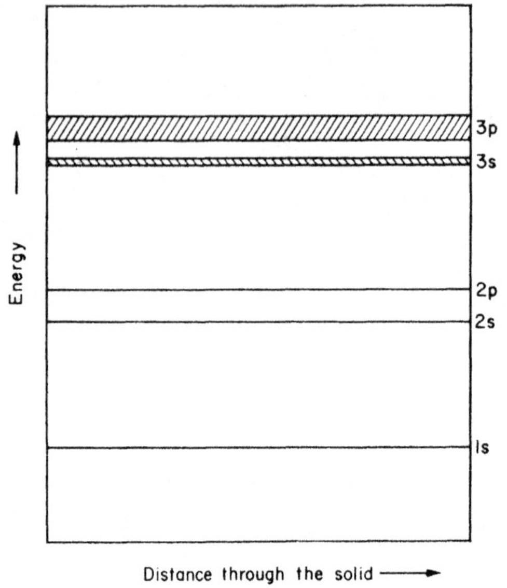

FIG. 11.4. Energy spectrum for a solid in which the interatomic spacing is a_3.

in this case (3*s*3*p*) hybrids. The second diagram, Fig. 11.3, corresponds to a narrow energy gap between the two bands, and the third diagram, Fig. 11.4, corresponds to a wide gap. In this context, narrow and wide are defined relative to the value of the thermal energy $k_B T$ at room temperature: in the first case the band gap E_g is comparable to $k_B T$; in the second E_g is very much greater than $k_B T$.

Now consider the introduction of electrons into these spectra. First, suppose that the solid consists of atoms having one $3s$ electron outside a neon shell. There are thus a total of N electrons to be accommodated. In the first case, Fig. 11.2, the quasi-continuum contains $8N$ states. At absolute zero the valence electrons will occupy the N lowest of these. Most of these will plainly be in states which are predominantly $3s$ in character, although some of the levels near the top of those occupied may contain some admixture of $3p$ character, depending on the nature and magnitude of the overlap. Above the highest filled state there is an empty quasi-continuum, so the valence electrons will behave very like a free fermion gas, and the material will resemble a Sommerfeld metal. It will conduct electricity and heat well, and it will have a small electronic specific heat. However, the agreement is only qualitative, because the quantitative predictions of Sommerfeld's theory depend on the use of a density of states function appropriate for a free particle in a uniform potential box. As stated here, the band theory is strictly qualitative, and gives no clue to the appropriate density of states function within the band. For both the other two spectra (Figs. 11.3 and 11.4) the band derived from the $3s$ states of the free atom is half full at absolute zero, the lowest N states being occupied. Again, above the highest filled state there extends an empty quasi-continuum: the fact that this quasi-continuum is less extensive in these cases than in the first is not important, since in fact the electrons at normal temperatures are excited into states only a few $k_B T$ above the Fermi energy, whereas in all these cases the empty band extends for some electron volts above the highest filled state. The only difference in the two situations is that in the latter two cases the electrons at the Fermi level are in pure s states, whereas in the first case they may possibly be in hybrid sp states: in other words, there is a likelihood of a difference in the form of the density of states curve, although again there is no way of assessing the difference quantitatively.

Now, consider the situation when the component atoms have *two* $3s$ electrons. In a solid having a band spectrum of the form of Fig. 11.2, the lowest $2N$ states in the composite band and filled, with a quasi-continuum of empty states extending above the Fermi surface. Now, irrespective of the form of the overlap, the electrons at the Fermi surface will be in hybrid (*sp*) states, and the material, although still a metallic conductor, can be expected to show considerable quantitative deviations from the Sommerfeld theory. This is in agreement with experiment: potassium, having one 3*s* electron is a metal which obeys quite well the Sommerfeld predictions, whereas calcium, with two 3*s* electrons in the free atom, is a metallic conductor in the solid state, but does not follow the Sommerfeld predictions at all well.

If the band spectrum is of the form shown in Fig. 11.4, with a large energy gap between the 3*s* and 3*p* bands, a solid having two 3*s* electrons per atom will have the lower band completely occupied, and the upper band completely empty. In order to increase its energy consequently an electron at the top of the lower band must acquire an amount of energy equal to the band gap energy. At low temperatures and low fields, this is extremely unlikely, recalling that

$$E_g \gg k_B T.$$

This material will consequently not exhibit electronic conduction at low temperatures: it will not have an electronic specific heat. As the temperature and/or the field is increased, a point will be reached at which the energy that can be acquired by the electron will equal the band gap energy: the conductivity will then abruptly increase. These properties are those of an insulator, which has an extremely poor conductivity at ordinary temperatures and fields, but exhibits "breakdown" at very high fields or temperatures, starting to conduct quite well. It follows that the conditions for a good insulator are:

(1) It must have sufficient electrons to fill exactly a band or bands, resulting in a spectrum consisting of wholly full and wholly empty bands. A necessary, but not sufficient, condition

for this is that the atoms or molecules composing the solid must have an *even* number of valence electrons.

(2) The gap between the highest energy state in the uppermost filled band and the lowest energy state in the lowest empty band (these may not be represented on the same band diagram, since they may not lie along the same direction) should be as large as possible, and anyway must be much greater than $k_B T$ at the highest desired working temperature.

If the band diagram has the form shown in Fig. 11.3, with the $3s$ and $3p$ bands separated by a gap that is comparable with $k_B T$, the situation is conceptually identical with that described above. The material is an insulator at absolute zero, with the lower band completely full, and the upper band completely empty. However, at room temperature the electrons can acquire from thermal sources—that is, as a result of collisions with phonons—energy comparable with the band gap energy, and thus electrons near the top of the lower band can be excited into the empty upper band. The material will thus be able to conduct electricity, and will have a conductivity that increases with temperature. Such a material is called an "intrinsic semiconductor", since its semiconducting properties are an intrinsic property of the pure material. The upper band is then called the conduction band, and the lower is called the valence band.

When an electron is ionised from the valence band into the conduction band as a result of a collision with a phonon, it leaves behind a vacant state at the top of the valence band. This vacant state now provides a means whereby electrons *in the valence band* can increase their energy, and so accelerate under the influence of an applied field and conduct electricity. This process may be easily visualised in a simple model. Suppose we have two totally enclosed tanks, one completely full of water, the other completely empty. If a potential field is now applied to the tanks by tilting them, there will be no net mass flow. However, if a small amount of water is transferred from the full to the empty tank, and the system is again tilted, there will be a net mass flow in both tanks: in the tank containing the small amount of water this will flow to the lowest part of the tank. In the nearly full tank also the water will flow to the lowest part of the tank; but an observer would find it easier to describe the motion by saying that a bubble (an absence of water) moved to the highest part of the tank. If this concept of a bubble was used for a quantitative description of the mass shift in the system, it would be necessary to ascribe to it an "effective negative mass" since it moves uphill, against gravity. It should be noted that it is not suggested that there exists a specific entity called a bubble having a negative mass, but merely that the motion of the water in the nearly full tank is such that it may be conveniently described in terms of a "pseudo-material"—a bubble which behaves *as though* it had negative mass.

In exactly the same way it is a great deal easier to describe the motion of the "hole" left in the valence band by the excited electron than it is to describe the motion of all the other electrons. Under the influence of a field an electron from lower in the band is excited to fill the vacant state at the top of the band; the appearance is consequently that as a result of the excitation the hole moves *down* (that is, to a position of lower energy) in the valence band. The energy increase experienced by an electron in a field depends on the product of the mass of the electron and its charge; consequently, the new quasi-particle, the hole, must have one or other (but not both) of these quantities reversed: it must have either positive charge or negative mass. There is no *a priori* reason at this point why one or other of these should be selected, but it will appear later that it is more appropriate to choose a positive mass and a positive charge for the hole.

The usefulness of this concept in the theory of semiconductors arises because of the symmetrical form of the Fermi–Dirac distribution function. If the distribution function for the electrons is f, then since a hole is the absence of an electron, the distribution function for holes must be $(1-f)$; but it was pointed out in Chapter 9 that the function $(1-f)$ is in fact the mirror image of the function f in the Fermi energy: if the direction in which energy is measured is simply reversed, taking the Fermi energy as zero, the two functions are identical. As a result of this, if one uses the concept of holes in the valence band and electrons in the conduction band as the current carriers, exactly the same functions may be used to describe the behaviour of both.

The hole is the second quasi-particle to be introduced; the first was the phonon.

In an intrinsic semiconductor the process of ionisation involves the production of an electron–hole pair, with the consumption of a quantum of energy:

$$\hbar\omega \geqq E_g \tag{11.1}$$

In the type of conduction described above, this quantum is provided by a phonon. However, the band gap energy may be so great that the quantum frequency is in the visible range, in which case ionisation may involve the absorption of a photon. Such a material exhibits photoconductivity: in the dark it is an insulator, because phonon energies are not sufficiently great to produce ionisation; but when illuminated the material conducts, as a result of photon induced ionisation. Initially, the electron–hole pair moves through the material as a pair, held together by Coulomb forces: in this (short-lived) state it is called an "exciton". Soon the weak bond between the two is broken by the field, and the electron and hole drift through the solid transporting charge. Eventually, however, the electron will come into close proximity with another hole: after a short period as an exciton they "recombine", emitting a quantum of energy equal to the band gap energy. The quantum absorbed in the initial ionisation process may be greater than the band gap energy, but the additional energy possessed by the carriers as a result of this soon decays, and recombination usually involves carriers at the extremes of the bands, so that the emitted quantum is very closely equal to the band gap energy. The result is that a photoconductor may absorb all radiation above a certain frequency, but the emission spectrum resulting from recombination will be nearly monochromatic. In some cases recombination may be delayed for some time: such a material is luminescent, absorbing light while illuminated, and re-emitting it later when the illumination is removed and the excited carriers slowly decay. Alternatively, recombination may be rapid: such a material is a phosphor, such as those used on TV screens: energetic electrons ionise the carriers, which decay to the band extrema, and then recombine, emitting a photon. This process occurs almost immediately in a good phosphor having little or no "afterglow". Actually, very few materials owe their semiconducting properties, their luminescent or phosphorescent properties to intrinsic processes.

The mean time between the ionisation of a carrier pair and their recombination is called the "mean carrier lifetime". The longer the mean carrier lifetime, the longer it is accelerating under the influence of an applied field, and consequently the greater the drift velocity, and hence conductivity. A high-conductivity semiconductor must therefore have a high carrier lifetime. Recombination is facilitated by such things as impurities, and defects in the regular crystalline array in the semiconductor, since these locally distort the band structure, bringing the band extrema closer together, and so making recombination easier.

Impurity Semiconductors

Most practical semiconductors owe their useful properties not to intrinsic ionisation across a small energy gap, but to the presence of small concentrations—typically one to ten parts per million—of impurities. This is partly because intrinsic semiconductors are rare, since the energy gap requirements are rather restrictive; but principally it is because the properties of an impurity semiconductor may be varied at will by varying the impurity concentration, while the properties of an intrinsic material are, of course, fixed. The role of impurities can be understood from the band scheme. Consider, for example, the element germanium, from group IV of the Periodic Table. This material has the diamond structure, with two atoms associated with each point of a face centred cubic lattice. Each lattice point has eight (*sp*) electrons associated with it, so the appropriate bands are full, and the material is an insulator at absolute zero. In fact, in the pure state it will continue to have very poor conductivity at room temperature, since the band gap is 0·70 eV, almost 30 times $k_B T$ at room temperature. To put these statements in another way: the germanium atoms are linked together by a tetrahedral arrangement of covalent bonds, formed as a result of the promotion of a 4*s* electron into a 4*p* state, followed by hybridisation of the states to form four equivalent 4(*sp*) orbitals. All the valence electrons are thus involved in the covalent bonds, and because they are localised they are incapable of drifting through the lattice carrying charge, until the energy of vibration of the lattice atoms becomes so great that some bonds may be broken. In a material in which the covalent bond structure is as strong as it is in germanium, this is clearly an unlikely event. When the bond is broken, the liberated electron drifts through the lattice transporting charge until it is captured by another incomplete bond. The original incomplete bond may capture one of the valence electrons from a neighbouring atom, thus itself moving through the lattice—this is the equivalent, in terms of this model, of the "hole" conduction discussed earlier.

Now suppose an atom from group V of the Periodic Table, such as arsenic or antimony, is introduced into the germanium lattice. This atom will enter into substitutional solid solution—that is, it will replace a germanium atom on the lattice. It has to contribute four electrons to complete the tetrahedral array of covalent bonds required by its immediate environment, but in addition to these it has one further valence electron. However, it also has one additional charge on the ion core, so the additional electron is bound to it by Coulomb attraction. The actual form of the potential field in which this extra electron moves is very complicated, composed of the Coulomb field of the additional charge on the impurity atom, the periodic field due to the lattice, and the field due to the electrons in the covalent bonds linking the structure. However, at very large distances, the discrete nature of the latter two potentials will not be evident, and the potential will then be a Coulomb field modified by the presence of the dielectric medium. As a limiting approximation consequently it is interesting to calculate the ground state energy of the hydrogen-like electron–ion core pair in a dielectric medium of dielectric constant ε. The method used will be that used in the calculation of the Bohr model. The potential energy of the pair at a separation r is

$$V = \frac{-e^2}{\varepsilon r}. \tag{11.2}$$

Using (2.30), the energy of the ground state ($n = 1$) becomes

$$E_D = \frac{e^4 m}{2\varepsilon^2 \hbar^2}$$

$$= \frac{13{\cdot}59935}{\varepsilon^2}\ \text{eV}. \tag{11.3}$$

For germanium, the dielectric constant is 15·8, so

$$E_D = 0{\cdot}049\ \text{eV} \tag{11.4}$$

and it is important to note that this energy does not depend on the nature of the impurity atom at all, so long as it is from group V of the Periodic Table: E_D should be the same for phosphorus, antimony, or arsenic. However, the value of E_D should be different for silicon, another group IV semiconductor, since the dielectric constant is different. This quantity of energy is now not dissimilar to that of a room temperature phonon, so ionisation of the additional electron away from the vicinity of the impurity atom should be relatively simple, and the material will consequently conduct by electron drift in the conduction band. Unlike the intrinsic case, there is no corresponding hole current, since the electron does not leave a vacant state in the valence band; it simply leaves a vacant state localised around the impurity atom, which is consequently not mobile. The charge carriers thus have a net negative sign, and so the material is said to be an "n-type semiconductor". The impurity atom is said to "donate" an electron to the conduction band, and is called a "donor" impurity: the quantity E_D is called the donor ionisation energy. The donor ionisation energy can be measured by measuring the temperature dependence of the conductivity: typical values are shown in Table 11.1. As required by the theory, the values are very similar for the different group V impurities. They are significantly less than the values calculated from (11.3), but this is scarcely surprising, in view of the grossly oversimplified potential used. It is interesting to calculate the radius of the orbit: (2.33) indicates that it will be

$$a_D = \frac{\hbar^2 \varepsilon}{me^2} = 0{\cdot}52947\varepsilon \text{ Å} \tag{11.5}$$

for germanium,

$$a_D = 8{\cdot}37 \text{ Å} \tag{11.6}$$

so that in fact the electron normally "sees" the donor charge through two or three germanium atoms. This accounts for the fact that the calculated values of the donor ionisation energy are as close to the experimental values as they are. One reason for the difference of the donor ionisation energy from donor to donor is because the donor atoms are of different size, and this causes a local perturbation in the germanium lattice which differs from donor to donor. However, there are many other reasons: this model is in fact a very crude representation of a very complicated situation. Since it is so easy to measure donor ionisation energies experimentally, the very considerable effort required to calculate them using a precise method is probably not justified.

On a band diagram, this situation may be described by a short line E_D below the bottom of the conduction band. The short line indicates that the unionised donor state is bound; its location indicates that a quantity of energy E_D is required to transfer the electron from the level to the conduction band. Often, an ionised donor level is indicated by a plus sign just below the bottom of the conduction band. Figure 11.5 shows a band diagram for an *n*-type semiconductor: for simplicity, the electron has been indicated by a small circle, but this is not usual.

TABLE 11.1. EXPERIMENTAL DONOR IONISATION ENERGIES COMPARED WITH VALUES CALCULATED USING EQUATION (11.3), IN eV

	ϵ	Calculated	P	As	Sb
Germanium	15·8	0·049	0·012	0·013	0·010
Silicon	11·7	0·099	0·045	0·050	0·039

The converse of the donor situation is observed when an impurity atom from group III enters into substitutional solid solution in germanium. Now, the atom has only three valence electrons, so one of the four covalent bonds surrounding it is in-

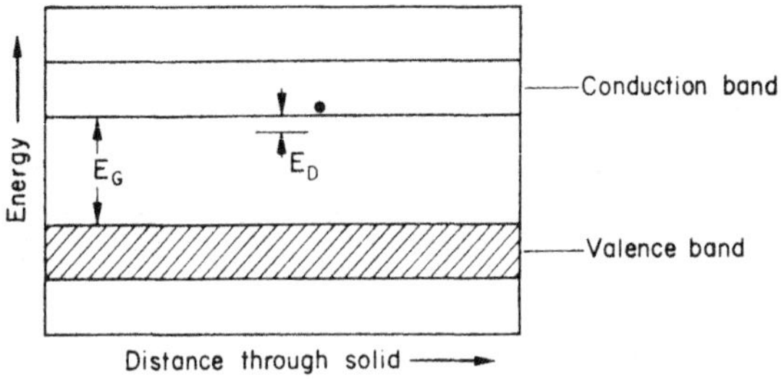

FIG. 11.5. Band diagram for an *n*-type semiconductor.

complete. This bond may be completed by an electron from a nearby atom, and the incomplete bond may thus drift away through the lattice, as described in the section on intrinsic semiconduction. The immediate vicinity of the impurity atom then has a net negative charge, and the immediate vicinity of the incomplete bond, which is now associated only with germanium atoms, has a net positive charge. There is thus a Coulomb attraction between the two. It is clear that this situation can be described as the ionisation of a hole away from the impurity atom, the positive hole being attracted towards the negatively

TABLE 11.2. EXPERIMENTAL VALUES OF ACCEPTOR IONISATION ENERGIES, eV

	Al	Ga	In
Germanium	0·010	0·011	0·011
Silicon	0·06	0·07	0·16

charged impurity ion. The situation is exactly analogous to the donor case considered above, except that now the impurity atom "accepts" an electron from the valence band as the result of the ionisation, the hole then conducting in the valence band.

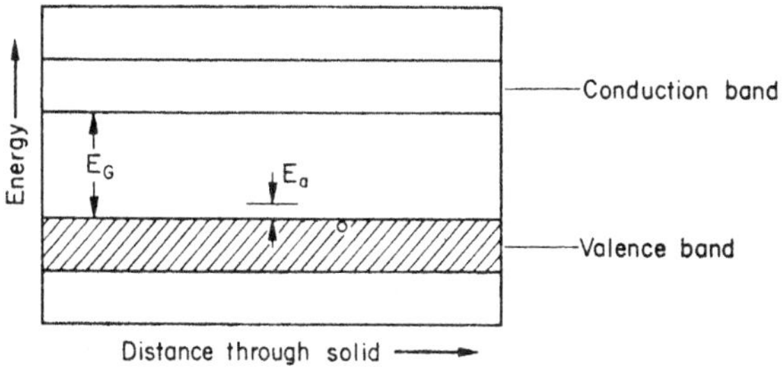

FIG. 11.6. Band diagram for a *p*-type semiconductor.

The charge carriers are thus positively charged, so the material is termed a "*p*-type semiconductor". The three-valent impurity is called an acceptor, and the ionisation energy is called the acceptor ionisation energy. If the same mass is allocated to the hole as to the electron, one would expect the acceptor ionisation energies to be very similar to the donor ionisation energies. Table 11.2 shows that this is approximately true.

This model demonstrates one of the reasons for a choice of a positive charge for the hole. In a band diagram, the acceptor levels are shown just above the top of the valence band, since the ionisation step involves the movement of an electron up from the top of the valence band. A band diagram for a *p*-type semiconductor is shown in Fig. 11.6.

A Summary of the Band Theory

Introduction of band theory has permitted a qualitative explanation of the existence of insulators, metals, intrinsic and impurity semiconductors. It predicts the phenomenon of insulator breakdown, and permits a semi-quantitative discussion of some aspects of impurity semiconductors. It accounts for the fact that the free electron theory works well for the monovalent metals, but fails to work for metals of higher valency. However, it has the serious disadvantage that it does not allow a calculation of a density of states curve within each band, and so does not permit a quantitative analysis of conduction. In addition, it shares with the free electron theory the disadvantage that it gives no clue as to the reason for the temperature dependent resistivity of metals.

Nearly Free Electron Theory

A more quantitative theory can be developed by modifying the free electron theory. Instead of using a uniform, time independent potential the periodic nature of the potential is recognised. Again, the one-electron approximation is assumed, and the potential is assumed to be time independent. The problem is then to calculate the eigenvalue spectrum of the Hamiltonian with a periodic potential. The analogous problem has been dealt with in Chapter 8 for elastic waves, and so long as the potential is conservative, the position dependent part of the separated Schrödinger equation is exactly the same as the elastic wave equation. Using the qualitative results of the Brillouin zone theory developed in Chapter 8, it follows that for long wavelengths (small **k**) the wave is unaware of the periodicity of the potential, behaving as though it were in a continuum. In terms of electrons, this means that the electrons behave as though they are free for small **k**, the wave function having the same form as in the free electron theory. At the walls of the zone the solution has the form of a standing wave, which one may regard as an incident wave and a reflected wave of equal intensity. In between these two limits, the solution may be regarded as a mixture of two terms, one a free electron wave, and one a reflected component, which increases in magnitude as the Brillouin zone boundary is approached. The definition of the Brillouin zone is exactly the same as in Chapter 8. In principle this description of the solution to the Schrödinger equation is true for any electron state, but in fact the core electrons remain for so long bound to one atom, tunnelling so infrequently through the potential barrier between the atoms, that the effective potential experienced by them is not periodic, but resembles more closely the free atom Coulomb potential. The effective potential is periodic only for electrons that are nearly free.

At the zone boundary, the wave function has the form of a standing wave. There are two possible standing waves, one with the nodes at the atom centres, and one with the nodes midway between the atoms. The electron density function $| \psi |^2$ for these two functions is different. In one case it indicates that the electron density is highest at the midpoint between the atoms, where the potential energy V is highest; in the other it indicates that the electron density is highest in the immediate vicinity of the atoms, where the potential energy is lowest. The kinetic energy is the same for both, since **k** is the same; so the total energy is higher in the first case. The E versus **k** curve is consequently double valued at the Brillouin zone wall. The free electron wave with the same wave number would give a uniform electron density, corresponding to a total energy between the two values corresponding to the standing waves. These various functions are sketched in Fig. 11.7.

In addition, the Brillouin zone is repeated periodically through **k**-space, and so the E versus **k** curve must also be repeated periodically through **k**-space. On the basis of these results, the general form of the E versus **k** curve according to the nearly free electron theory is sketched in Fig. 11.8a. The dotted line shows

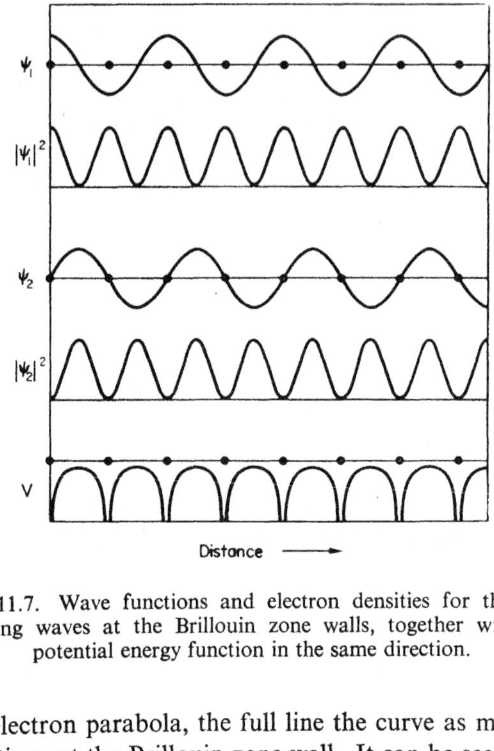

FIG. 11.7. Wave functions and electron densities for the two standing waves at the Brillouin zone walls, together with the potential energy function in the same direction.

the free electron parabola, the full line the curve as modified by the reflections at the Brillouin zone wall. It can be seen that this theory leads to a band energy spectrum, exactly as above; and within the band the energy is quasi-continuous. The quasi-continuity of the energy arises when the finite size of the specimen is taken into account [equation (8.61)] with exactly the same number of states per atom as the qualitative argument in the first part of this chapter concluded. In Fig. 11.8b the same curves are shown, periodically repeated through **k**-space.

From these curves some qualitative conclusions may be drawn.

(1) The curves do not deviate from the free electron form until

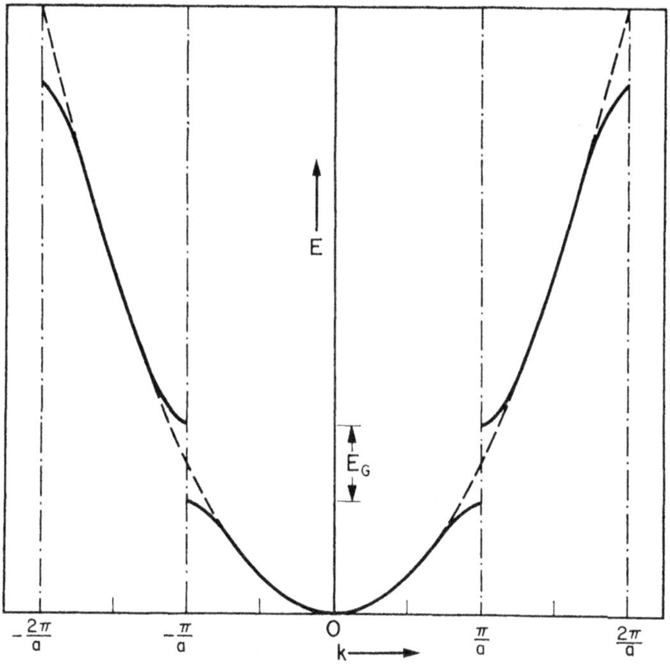

FIG. 11.8a. The energy–wave number curve for the nearly free electron theory (full line) compared with the free electron parabola (dotted line).

the wave number approaches the zone wall. In terms of the vocabulary of Chapter 10, the Fermi surface will remain approximately spherical until it approaches the zone walls, and the results of the free electron theory will be quantitatively correct. This is true for the alkali metals and the noble metals of group I which have only one valence electron per atom, and so only half fill the first band. In fact, experiment shows that even in these materials the Fermi surface is appreciably distorted from spherical, but the distortion does not importantly modify the quantitative results of the free electron theory.

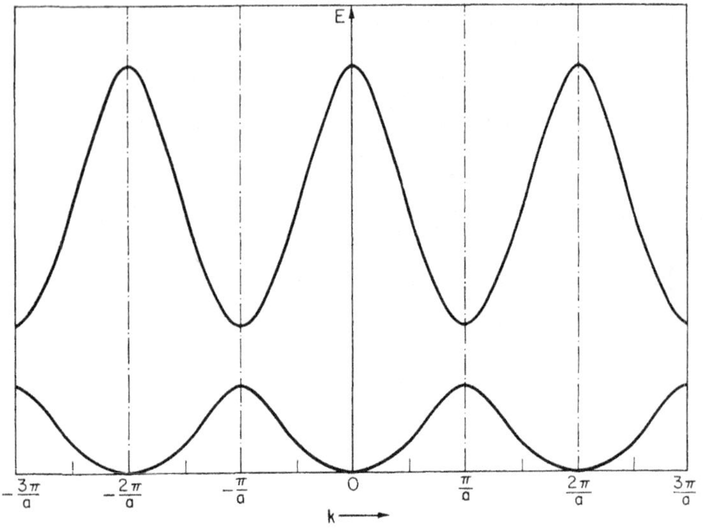

FIG. 11.8b. The energy–wave number curve for the nearly free electron theory periodically repeated through **k**-space.

(2) The shape of the E versus **k** curve at the bottom of the second band is *qualitatively* similar to that at the bottom of the first band. It may be tentatively concluded that carriers at the bottom of any band will behave in a manner very similar to free electrons.

(3) An even more interesting point is that the curve at the *top* of the first band is also similar to that at the bottom of the band, provided the energy is reversed, being measured from the top of the band down. This clearly indicates that the states at the top of a band may be regarded as "hole" states; and that furthermore

because of the qualitative similarity in the curves the holes may be expected to behave in a manner very similar to free electrons. It is this peculiarity of the E versus **k** curves that makes the concept of holes so useful.

The Concept of Effective Mass

It is interesting to consider at this point the dynamics of electron or hole movement in a crystal. If the carrier is moving in a narrow band, with consequently a high density of states, it follows that the atomic states from which the band is derived do not overlap very much. Consequently, an electron or hole spends most of its time performing "atom-like" orbitals around a particular atom, only infrequently migrating to a neighbouring atom. Even if vacant states are available in the band, the charge carrier consequently will accelerate only slowly under the influence of an applied field. It will therefore *appear* to have mass much larger than it really has, in terms of a free electron approach. Conversely, if the density of states is very low a single excitation significantly increases the energy of the carrier, so it accelerates *more rapidly* than would a free carrier having a mass of an electron: it consequently appears to have a lower mass. From this, it appears that it may be possible to describe the very complicated motion of a carrier in the lattice in terms of free electron dynamics, by simply treating the carriers as though they had masses different from that of the free electron. If the acceleration under the influence of an applied force F is A, then in terms of free particle dynamics the particle behaves as though it had a mass

$$m^* = F/A \tag{11.7}$$

and the quantity m^* is called the "effective mass" of the carrier.

Suppose that the carrier (electron or hole) can be regarded as a wave packet made up of states near a particular value k in the band. The group velocity is then

$$v_g = \frac{d\omega}{dk} \tag{3.12}$$

and

$$\omega = \frac{E}{\hbar} \tag{1.1}$$

so

$$v_g = \frac{1}{\hbar}\frac{dE}{dk}. \tag{11.8}$$

Now an electric field $\mathscr{E}$ is applied to the solid, which in a time δt increases the energy by an amount δE, where

$$\delta E = e\mathscr{E}v_g\delta t. \tag{11.9}$$

But

$$\delta E = \frac{dE}{dk}\,\delta k$$

$$= \hbar v_g \delta k \tag{11.10}$$

from (11.8). Inserting this value in (11.9),

$$\delta k = \frac{e\mathscr{E}}{\hbar}\,\delta t$$

or,

$$\hbar\frac{dk}{dt} = e\mathscr{E}. \tag{11.11}$$

The right-hand side of this expression is the external force on the electron, which in free space would be equal to the mass times the acceleration, mdv/dt. Now from (11.6)

$$\frac{dv_g}{dt} = \frac{1}{\hbar}\frac{d^2E}{dkdt} = \frac{1}{\hbar}\frac{d^2E}{dk^2}\frac{dk}{dt}$$

or,

$$\hbar\frac{dk}{dt} = \frac{\hbar^2}{(d^2E/dk^2)}\frac{dv_g}{dt}. \tag{11.12}$$

Comparing (11.11) and (11.12), it appears that the effective mass of a carrier in a solid is

$$m^* = \hbar^2\left(\frac{d^2E}{dk^2}\right)^{-1}. \tag{11.13}$$

Plainly, the concept of an "effective mass" is only useful if the acceleration of the particle is more or less linear with the applied force, which means that m^* must be reasonably constant. For any part of the band for which the E versus **k** curve has the form

$$E = E_0 \pm \text{constant}.k^2 \qquad (11.14)$$

the effective mass is constant. It should be noticed that the E versus **k** curve at the top of the lower band in Fig. 11.8b has the form of an inverted parabola, so the electron effective mass is negative; however, the hole effective mass is positive, since for holes the energy is measured from the top downwards. This is the formal justification for the choice of a positive mass and a positive charge for a hole. An interesting point is that at the point of inflexion of the E versus **k** curves the effective mass of the carriers, electrons or holes, becomes zero.

The E versus **k** curve is a function of direction in **k**-space. The curve may be everywhere parabolic, but the shape of the parabola may vary. In this case, the effective mass at any point in **k**-space contains three terms, one referring to each direction. These three terms may be different at every point in **k**-space, without the concept of effective mass losing its usefulness: a quantity like this is termed a tensor: the scalar mass of the free particle is replaced by a tensor effective mass in the crystal. Fortunately, in many real problems the effective mass of the carriers is constant over the Fermi surface.

It is difficult to visualise the meaning of a negative effective mass. While it may be easy to justify an effective mass which is larger or smaller than the real mass of the carrier, in terms of acceleration being more or less difficult in the crystal than in free space, it is difficult to see how the carrier can move in the *opposite* direction to that suggested by the free electron approach. This, however, arises because the change in curvature of the E versus **k** curve which produces the negative effective mass is caused by the increasing contribution of a wave *reflected* by the lattice. Under the influence of a field although the electron tries to move in the same direction as suggested by free electron dynamics, it is reflected by the lattice so strongly that the net movement is in the opposite direction, hence the negative effective mass.

It appears, therefore, that it is often possible to replace the complicated situation of the valence electrons in the lattice by the much simpler model of a free fermion gas of electrons or holes, with appropriately chosen effective masses. Where this identification can be made, the quantitative results of the free electron theory, and of free electron dynamics may be incorporated directly into the band theory, simply replacing m and e by m^* and $\pm e$, depending on the sign of the carrier. When this simplifying substitution is made, the "electrons" of which one speaks are as much quasi-particles as holes and phonons.

Energy Contours in the Brillouin Zone

The E versus $\mathbf{k}$ curve varies with the direction in $\mathbf{k}$-space, and consequently the constant energy surfaces in $\mathbf{k}$-space will no longer be spherical. At the origin of $\mathbf{k}$-space, where the wave functions are not significantly modified by the reflected component, the E versus $\mathbf{k}$ curve is still parabolic, and is the same for every direction in $\mathbf{k}$-space, so that near the origin the constant energy surfaces retain their spherical form. As an example, consider a two-dimensional crystal in which the atoms are arranged in a square lattice. The Brillouin zone walls in the appropriate two-dimensional $\mathbf{k}$-space are consequently defined by the four lines

$$\left.\begin{aligned} k_x &= \pm\frac{\pi}{a} \\ k_y &= \pm\frac{\pi}{a} \end{aligned}\right\} \qquad (11.15)$$

so that the Brillouin zone is also square. The shortest distance from the origin to a zone wall is along one of the axes: the longest is along one of the diagonals. Typical E versus $\mathbf{k}$ curves for these two extremes are sketched in Fig. 11.9. From these curves, the energy contours in the zone are sketched in Fig. 11.10. The contours depart from the circular first along the axes, the curve bulging towards the zone walls, since plainly the distortion from the free electron curve is such that a given E

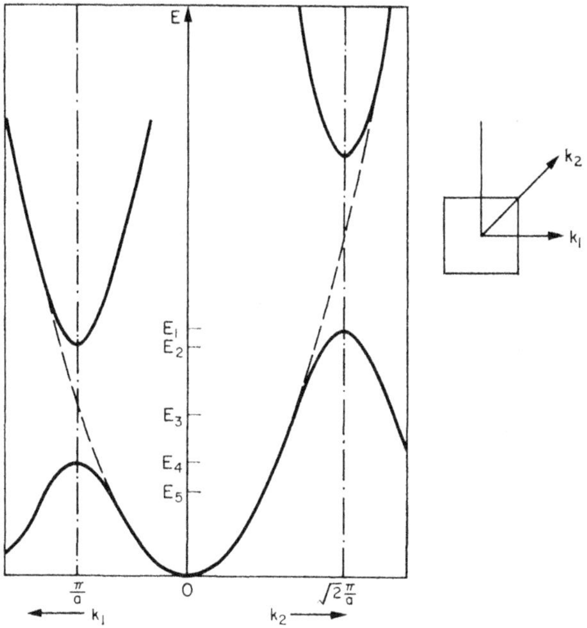

FIG. 11.9. Energy versus wave number curves for two directions in a square Brillouin zone. (Free electron parabola shown dotted.)

corresponds to a *larger* value of **k**. Eventually, the contour touches the zone walls. Now, since the zone is repeated periodically in **k**-space, and since the E versus **k** curves must be smooth curves, without discontinuities in slope, it follows that the contours must intersect the zone boundaries normally. For the example sketched, the second band will start to fill before the first is completely full, since the highest energy level in the first band (along the zone diagonal) is higher than the lowest energy level in the second band (along the axis). In this case, consequently, there is band overlap, although the energy gap at the zone walls never vanishes. Plainly, this may not be so: there may

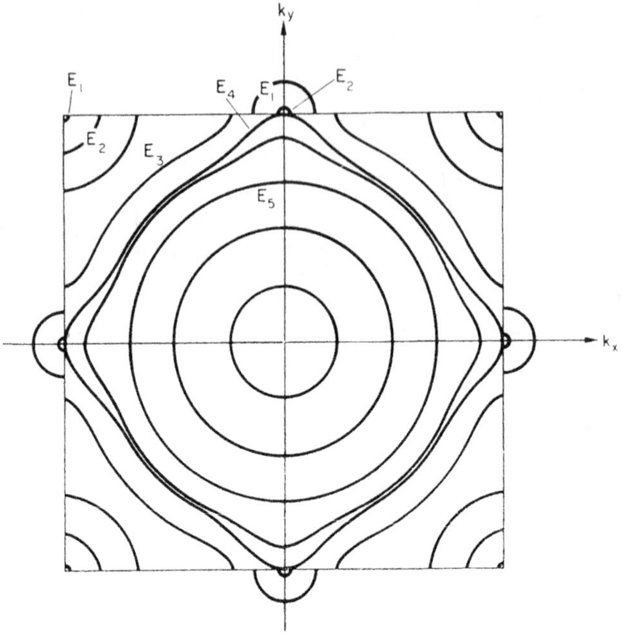

FIG. 11.10. Energy contours in a square Brillouin zone.

be a band gap in the sense that the second band starts to fill only after the first is fully occupied.

Consider now the case where a material with energy contours of the form shown in Fig. 11.10 has the first band nearly full. The zone is shown repeated in **k**-space in Fig. 11.11, with the occupied states shaded. It can be seen that the unoccupied states are surrounded by circles, exactly as the filled (free electron) states near the centre of **k**-space are surrounded by circles. This demonstrates yet again that the empty states at the top of a band may be represented as free hole states. In the three-dimensional case, Fermi surfaces are often very complicated, with very considerable variations of curvature. It is customary to refer to "hole states" and "electron states" on the same

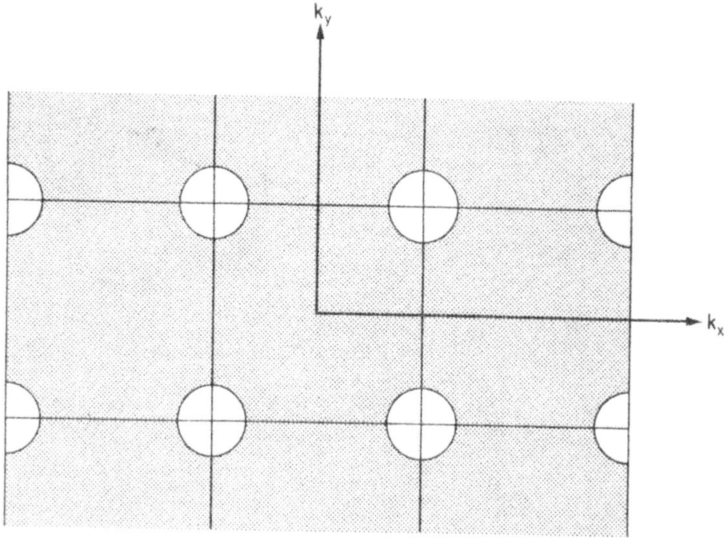

FIG. 11.11. Nearly full zone repeated periodically in **k**-space. Filled states are shaded.

surface, depending on whether the curvature is negative or positive with respect to the origin of **k**-space. It is doubtful, however, whether the concept of quasi-free electrons and holes is useful for the complicated multi-sheet Fermi surfaces typical of polyvalent metals, but in semiconductors, where in the main one is dealing with very low concentrations of carriers, the concepts of free quasi-electrons and free holes are extremely useful.

Density of States Functions in Band Theory

From the sketches drawn above, the density of states function can be drawn. At low energies, the situation does not deviate significantly from the free electron case, so that the density of states curve is parabolic at the bottom of each band, although it must be remembered that the equation of the parabola is not

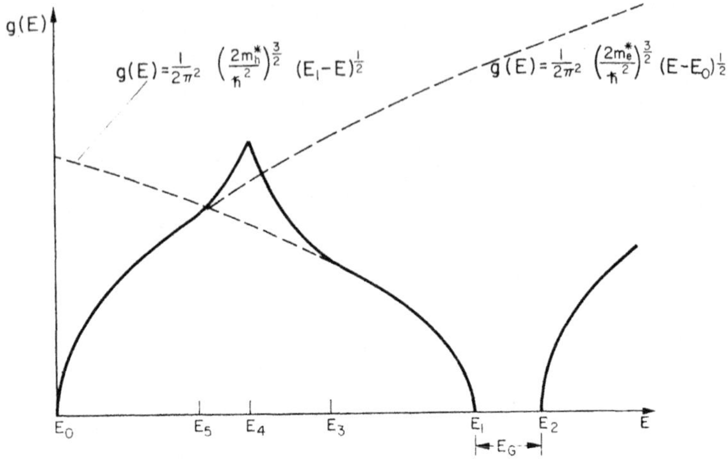

FIG. 11.12. Density of states curve in the Brillouin zone. In the example shown, there is no band overlap: otherwise the electron mass m_e*, the hole mass m_h* and various energies have been chosen to be compatible with Fig. 11.9.

exactly the same as (9.46): the electron mass m must be replaced by the effective mass m_e^*, and the energy E must be measured from the bottom of the band, and so becomes $E-E_0$, where E_0 is the energy of the bottom of the band. As the energy contours start to bulge towards the zone walls, each elementary energy range contains successively *more* states than the spherical shells of the free electron theory: the density of states curve, consequently, deviates positively from the free electron parabola, reaching a maximum as the contours touch the zone walls. After this point, the density of states falls, since only that part of the elementary shells which fall inside the Brillouin zone contribute. Eventually, as the surfaces of constant energy become inverted spheres enclosing free hole states the density of states curve again becomes parabolic, but in the reverse sense,

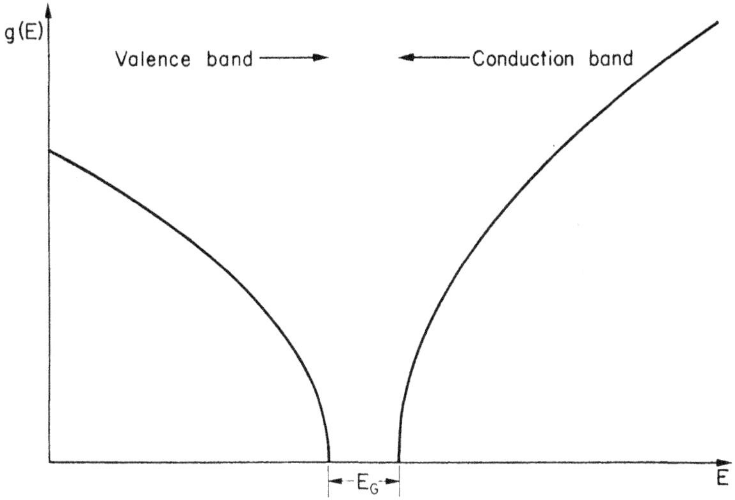

FIG. 11.13. Density of states curve for a semiconductor having two bands of standard form, with $m_h^* < m_e^*$.

with m in (9.46) replaced by m_h^*, the effective hole mass, and E replaced by $E_1 - E$, where E_1 is the energy at the top of the band. The density of states function for two non-overlapping bands is shown in Fig. 11.12. Actually, of course, the density of states function is only parabolic at the band extrema if the "free quasi-particle" model is valid: frequently, the function is much more complicated than this. In addition, the density of states curve may become infinite at points in the zone (provided the area beneath the curve remains finite), and for a zone of complicated shape it may have a number of points where the slope is discontinuous. In addition, detailed features of the curve depends on the interaction between the Fermi surface and the Brillouin

TABLE 11.3. SOME CARRIER EFFECTIVE MASSES

	m_e^*/m_0	m_h^*/m_0
Elementary Semiconductors		
Ge	0·8	0·04; 0·3
Si	0·6	0·16; 0·5
α Sn	0·15	0·2
III—V Semiconductors		
Al Sb	0·3	1·8
Ga As	0·07	0·5
Ga Sb	0·2	0·4
In P	0·07	0·4
In As	0·05	0·33
In Sb	0·01	0·2
Other Compound Semiconductors		
Si C	0·8	1·2
Cd Te	0·14	0·37
Pb S	0·22	0·1
Mg_2Ge	6·6	3·5
Mg_3Sb_2	0·32	0·57
Bi_2Te_3	0·45	0·51

All these values are very approximate, since different measuring techniques often give different values. Further, the effective mass is in general a tensor quantity which cannot be defined by a single number. Finally, there may be more than one carrier mass in a material. Thus, the two masses given for holes in germanium and silicon correspond to "light" and "heavy" holes, respectively.

zone walls, which in turn depends on the magnitude of the energy discontinuity at the walls. This discontinuity is itself determined by the nature of the interatomic interactions.

A band having a density of states curve which at the bottom of the band has the form

$$g(E) = \frac{1}{2\pi^2}\left(\frac{2m_e^*}{\hbar^2}\right)^{\frac{3}{2}}(E-E_0)^{\frac{1}{2}} \tag{11.16}$$

is called a band of standard form. If the top of the band has a parabolic density of states curve:

$$g(E) = \frac{1}{2\pi^2}\left(\frac{2m_h^*}{\hbar^2}\right)^{\frac{3}{2}}(E_1-E)^{\frac{1}{2}} \tag{11.17}$$

it is said to be an inverted band of standard form. In these expressions, E_0 and E_1 are the energies of the bottom and top of the band respectively; m_e^* is the effective mass of an electron at the bottom of the band, and m^* is the effective mass of a hole at the top of the band. The simplest semiconductor has a valence band of inverted standard form and a conduction band of standard form. The density of states curve for such a material is shown in Fig. 11.13. Notice that in Fig. 11.13 the magnitude of $g(E)$ for a given energy interval from the band extremum is smaller for the valence band than for the conduction band, implying that

$$m_h^* < m_e^*. \tag{11.18}$$

Some typical effective masses are listed in Table 11.3.

When the bands overlap, the density of states function may become very complicated: even for a simple overlap of two bands of similar form to those sketched in Fig. 11.12, the resultant curve is far from simple, as shown in Fig. 11.14.

In this chapter a band theory has first of all been developed from simple qualitative arguments by constructing a solid from free atoms, and it has been demonstrated that a number of interesting qualitative statements can be made on the basis of this simple model. In the second part of the chapter a more quantitative theory has been developed, starting from the opposite extreme of a free electron model. This leads to a theory which retains all the qualitative features of the first approach, but in addition permits some semiquantitative statements to be made, and suggests a form of the density of states curve which can be

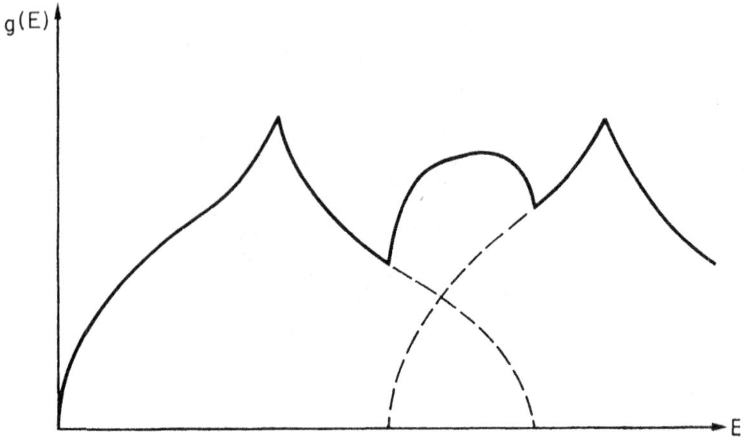

FIG. 11.14. Density of states curve for two overlapping bands.

related to the free electron density of states curve by means of a simple parameter, the effective carrier mass, which can be experimentally determined. So long as this substitution is justified, the quantitative results of the free electron theory may be directly incorporated into a theory of real solids.

CHAPTER 12

Some Results of Band Theory

IN THE main, quantitative calculations based on the band theory of semiconductors will be dealt with by Dr. Leck in his book *Theory of Semiconductor Junction Devices.* However, in order to demonstrate some of the principles a few relations will be derived in this chapter.

The conductivity of a semiconductor depends on the concentration of charge carriers of both sorts and on their drift velocity. This latter quantity is usually defined in terms of the "carrier mobility" which is the drift velocity under a unit applied field. In practical units therefore the dimensions of carrier mobility are $cm^2\ sec^{-1}\ V^{-1}$. Under a given applied field the drift velocities of holes and electrons are in opposite directions, and consequently have opposite signs. Conventionally, however, mobilities are always defined as positive. Writing n and p for the concentrations of electrons and holes respectively, and λ_e and λ_h for the mobilities, the conductivity may be written

$$\sigma = |\, e \,|(n\lambda_e + p\lambda_h). \tag{12.1}$$

In order to calculate the conductivity of a semiconductor the magnitudes of all four of the unknowns on the right-hand side of (12.1) must be computed. In fact, the calculation of carrier mobilities is difficult and rather beyond the scope of the present book. Seitz has calculated that for covalent semiconductors such as germanium or silicon

$$\lambda \propto m^{*-\frac{5}{2}}\, T^{-\frac{3}{2}}. \tag{12.2}$$

The carrier concentrations may, however, be calculated much

more simply. It will be assumed that both the valence band and the conduction band are of standard form, characterised by a simple scalar effective mass. For further simplicity, it will be assumed that the temperature is low, so that

$$E_g \gg k_B T. \tag{12.3}$$

In this case the carrier concentration will be small, so that the Fermi–Dirac distribution function takes the non-degenerate form

$$f(E) \simeq \exp\left(\frac{\mu_F - E}{k_B T}\right) \tag{12.4}$$

where as before μ_F is the Fermi energy. In fact, most practical semiconductors are non-degenerate; a notable exception is indium antimonide, for which the energy gap is only 0·18 eV.

The density of states function in the conduction band may be written

$$g_e(E) = \frac{1}{2\pi^2}\left(\frac{2m_e^*}{\hbar^2}\right)^{\frac{3}{2}}(E - E_g)^{\frac{1}{2}} \tag{12.5}$$

where the energy E is measured from the top of the valence band. Now,

$$n = \int_{E_g}^{\infty} g_e(E) f(E) dE \tag{12.6}$$

and using (12.4) and (12.5) this becomes

$$n = \frac{1}{2\pi^2}\left(\frac{2m_e^*}{\hbar^2}\right)^{\frac{3}{2}} \exp\left(\frac{\mu_F}{k_B T}\right)\int_{E_g}^{\infty}(E - E_g)^{\frac{1}{2}} \exp\left(\frac{E}{k_B T}\right) dE \tag{12.7}$$

which integrates to

$$n = 2\left(\frac{2\pi m_e^* k_B T}{\hbar^2}\right)^{\frac{3}{2}} \exp\left(\frac{\mu_F - E_g}{k_B T}\right). \tag{12.8}$$

As pointed out in Chapter 11, the distribution functions for holes is plainly

$$f_h = 1 - f_e \tag{12.9}$$

where f_e is the distribution function for electrons, since the presence of a hole is the absence of an electron. This function is a mirror image in the Fermi energy of the function f_e:

$$f_h = 1 - \frac{1}{\exp[(E-\mu_F)/k_B T]+1} = \frac{1}{\exp[(\mu_F - E)/k_B T]+1} \tag{12.10}$$

and in the non-degenerate limit this becomes

$$f_h = \exp\left(\frac{E-\mu_F}{k_B T}\right). \tag{12.11}$$

The density of hole states is

$$g_h(E) = \frac{1}{2\pi^2}\left(\frac{2m_h^*}{\hbar^2}\right)^{\frac{3}{2}}(-E)^{\frac{1}{2}} \tag{12.12}$$

again measuring the energy from the top of the valence band, and so

$$p = \int_{-\infty}^{0} g_h(E) f_h(E) dE$$

$$= 2\left(\frac{2\pi m_h^* k_B T}{\hbar^2}\right)^{\frac{3}{2}} \exp\left(-\frac{\mu_F}{k_B T}\right). \tag{12.13}$$

Multiplying (12.8) and (12.13),

$$np = 4\left(\frac{2\pi k_B T}{\hbar^2}\right)^3 (m_e^* m_h^*)^{\frac{3}{2}} \exp\left(-\frac{E_g}{k_B T}\right). \tag{12.14}$$

This means that the product of the concentrations of electrons and holes is a constant for a given material at a given temperature. The presence of small amounts of impurities does not affect the validity of (12.14), nor does it affect the numerical value of the product. Consequently, if donor impurities are added to a material the concentration of electrons increases and the concentration of holes falls in such a way as to keep the concentration product constant. It follows that an n-type semiconductor always has a certain amount of hole conduction, and a p-type semiconductor has a certain amount of electron conduction. These carriers are called minority carriers: their presence is very important in the operation of semiconductor devices.

For an intrinsic material, the concentration of electrons always equals the concentration of holes:

$$n_i = p_i \tag{12.15}$$

and from (12.14)

$$n_i = 2\left(\frac{2\pi k_B T}{\hbar^2}\right)^{\frac{3}{2}} (m_e^* m_h^*)^{\frac{3}{4}} \exp\left(-\frac{E_g}{2k_B T}\right) . \tag{12.16}$$

Combining (12.1), (12.2) and (12.16) it follows that the conductivity of an intrinsic semiconductor may be written

$$\sigma = \text{constant}\,.\,T^{-\frac{3}{2}} \exp\left(-\frac{E_g}{2k_B T}\right) \tag{12.17}$$

or

$$\ln \sigma = A - \tfrac{3}{2} \ln T - \frac{E_g}{2k_B T} . \tag{12.18}$$

The last term on the right-hand side varies much more rapidly with temperature than the second, so a graph of the logarithm of the conductivity versus the reciprocal of the absolute temperature is very nearly rectilinear: from the slope of this line the value of the energy gap for the material may be directly calculated. Although at room temperature the majority of semiconductors are impurity conductors, with the intrinsic mechanism swamped by the impurity contribution, as the temperature is raised the concentration of intrinsic carriers increases while the concentration of impurity carriers varies only slightly, since nearly all the impurity atoms will be ionised at room temperature. If the temperature is high enough consequently, n approaches p: the material is then said to be in its intrinsic range. In the intrinsic range (12.18) holds; below this range (12.18) does not hold. It is consequently quite easy to see from the experimental results when the material "becomes intrinsic", and from the slope of the graph at temperatures above this point to determine the band gap. Some typical band gap energies and carrier mobilities are shown in Table 12.1.

A further point of interest that emerges from the preceding equations is that, from (12.8) and (12.13),

$$\mu_F = \frac{3k_BT}{4}\ln\left(\frac{m_h^*}{m_e^*}\right) + \frac{E_g}{2} \tag{12.19}$$

TABLE 12.1

Material	Band gap (eV)	Electron mobility ($cm^2\ sec^{-1}\ V^{-1}$)	Hole mobility ($cm^2\ sec^{-1}\ V^{-1}$)
Silicon	1·10	1600	400
Germanium	0·7	3800	1800
Diamond	6	1800	1200
InSb	0·18	77000	1250
GaAs	1·4		
PbTe	0·30	17000	
CdS	2·42		
Cu_2O	2·1		

Compare: k_BT at room temperature ~ 0·026 eV; energy of a visible photon ~ 2·5 eV; mobility of an electron in metallic copper at room temperature 35 $cm^2\ sec^{-1}\ V^{-1}$.

and, in particular, if $m_e^* = m_h^*$, the Fermi level lies in the centre of the forbidden energy gap, with a position that is independent of temperature.

Thermal Ionisation of Impurity Atoms

Now consider the case of a material containing acceptor levels at an energy E_a and donor levels at an energy E_c, both energies being measured from the top of the valence band. In terms of the notation used in Chapter 11, $E_c = E_g - E_d$. The concentrations of donor and acceptor atoms are N_d and N_a per unit volume respectively: of these a concentration N_d^0 and N_a^0 are un-ionised, and a concentration N_d' and N_a' are ionised. Then

$$\left.\begin{aligned} N_d' &= N_d - N_d^0 \\ N_a' &= N_a - N_a^0 \end{aligned}\right\} \tag{12.20}$$

Using the Fermi–Dirac distribution function for electrons, the fraction of filled donor states (that is, un-ionised donor states) is

$$\frac{N_d^0}{N_d} = \frac{1}{1 - \exp\left[(E_c - \mu_F)/k_B T\right]} \tag{12.21}$$

and the fraction of filled acceptor states (that is, *ionised* acceptor states) is

$$\frac{N_a'}{N_a} = \frac{1}{1 + \exp\left[(E_a - \mu_F)/k_B T\right]}\,. \tag{12.22}$$

The system as a whole must be electrically neutral, so that

$$N_d' + p = N_a' + n \tag{12.23}$$

where n and p are the concentrations of electrons in the conduction band and holes in the valence band, respectively. Combining (12.8), (12.13), (12.21), (12.22) and (12.23),

$$2\left(\frac{2\pi m_h^* k_B T}{h^2}\right)^{\frac{3}{2}} \exp\left(-\frac{\mu_F}{k_B T}\right) + \frac{N_d}{1 + \exp\left[-(E_c - \mu_F)/k_B T\right]}$$
$$= 2\left(\frac{2\pi m_e^* k_B T}{h^2}\right)^{\frac{3}{2}} \exp\left(\frac{\mu_F - E_g}{k_B T}\right) + \frac{N_a}{1 + \exp\left[(E_a - \mu_F)/k_B T\right]}\,. \tag{12.24}$$

As it stands (12.24) is very complicated. However, for a real n-type semiconductor sufficient donor impurity has been added that by comparison N_a may be set equal to zero. In addition, remembering (12.14), the concentration of minority carrier may be neglected at low temperatures: this is equivalent to saying that for these materials

$$\exp\left[(\mu_F - E_g)/k_B T\right] \gg \exp\left(-\mu_F/k_B T\right) \tag{12.25}$$

which indicates that the Fermi level is either much nearer the conduction band than the valence band or indeed lies within the conduction band, with $\mu_F > E_g$. In this case (12.24) becomes

$$\frac{N_d}{1+\exp\left[-(E_c-\mu_F)/k_B T\right]} = 2\left(\frac{2\pi m_e^* k_B T}{h^2}\right)^{\frac{3}{2}} \exp\left(\frac{\mu_F - E_g}{k_B T}\right). \tag{12.26}$$

This can be rewritten as a quadratic equation in $\exp(\mu_F/k_B T)$:

$$\left\{2\left(\frac{2\pi m_e^* k_B T}{h^2}\right)\exp\left[\frac{-(E_c+E_g)}{k_B T}\right]\right\}\exp\left(\frac{2\mu_F}{k_B T}\right)$$
$$+\left[2\left(\frac{2\pi m_e^* k_B T}{h^2}\right)^{\frac{3}{2}}\exp\left(-\frac{E_g}{k_B T}\right)\right]\exp\left(\frac{\mu_F}{k_B T}\right) - N_d = 0 \tag{12.27}$$

for which the solution is

$$\exp\left(\frac{\mu_F}{k_B T}\right) = \frac{-1+\{1+4[N_d/2\,(2\pi m_e^* k_B T/h^2)^{\frac{3}{2}}]\exp(E_g-E_c)/k_B T\}^{\frac{1}{2}}}{2\exp(-E_c/k_B T)}. \tag{12.28}$$

This can be evaluated in two limiting situations. First, if the temperature is high, or the donor concentration is low (recalling that in fact neither of these can be permitted to become so extreme that the minority carrier concentration becomes significant) then

$$4\left(\frac{N_d}{2(2\pi m_e^* k_B T/h^2)^{\frac{3}{2}}}\right)\exp(E_g-E_c)/k_B T \ll 1 \tag{12.29}$$

and

$$\exp\left(\frac{\mu_F}{k_B T}\right) \simeq \frac{N_d}{2(2\pi m_e^* k_B T/h^2)^{\frac{3}{2}}}\exp\left(\frac{E_g}{k_B T}\right) \tag{12.30}$$

so that the Fermi energy is

$$\mu_F \simeq E_g + k_B T \ln \left[\frac{N_d}{2(2\pi m_e^* k_B T/h^2)^{\frac{3}{2}}} \right] \quad (12.31)$$

and the electron concentration in the conduction band is

$$n = 2\left(\frac{2\pi m_e^* k_B T}{h^2}\right)^{\frac{3}{2}} \exp\left(\frac{\mu_F - E_g}{k_B T}\right) \simeq N_d \quad (12.32)$$

indicating that in fact under these conditions all the donors will be ionised. The opposite extreme is that of very large donor concentration or very low temperature. In this case

$$4\left[\frac{N_d}{2(2\pi m_e^* k_B T/h^2)^{\frac{3}{2}}}\right] \exp\left(\frac{E_g - E_c}{k_B T}\right) \gg 1 \quad (12.33)$$

and

$$\exp\left(\frac{\mu_F}{k_B T}\right) \simeq \left[\frac{N_d}{2(2\pi m_e^* k_B T/h^2)^{\frac{3}{2}}}\right]^{\frac{1}{2}} \exp\left(\frac{E_g + E_c}{2kT}\right) \quad (12.34)$$

giving the Fermi energy as

$$\mu_F = \tfrac{1}{2}(E_g + E_c) + \frac{k_B T}{2} \ln \left[\frac{N_d}{2(2\pi m_e^* k_B T/h^2)^{\frac{3}{2}}}\right] \quad (12.35)$$

and the electron concentration in the conduction band as

$$n = (2N_d)^{\frac{1}{2}} \left(\frac{2\pi m_e^* k_B T}{h^2}\right)^{\frac{3}{4}} \exp\left(\frac{E_d}{2k_B T}\right) \quad (12.36)$$

where E_d, the donor ionisation energy, has been written for $E_g - E_c$. Clearly, (12.36) corresponds to a Maxwell–Boltzmann limit of the Fermi–Dirac distribution law, with a simple exponential ionisation of donors. Unlike the similar expression for the intrinsic concentration (12.36) does not indicate that the donor ionisation energy can be measured by determining the temperature dependence of the conductivity of an n-type semiconductor at low temperatures. This is because in the limits where (12.33) is valid either the donor concentration is so high that the donor states overlap, giving rise to an "impurity band", or else the temperature is so low that the concentration of electrons becomes comparable to the concentration of acceptor atoms always present as an unavoidable impurity.

Conductivity of Metals

It appears from the discussion in Chapter 11 that near the centre of a zone the energy–wave number relationship is qualitatively the same as that for free electrons. In general, an effective mass may be defined at each point of the Fermi surface by a relationship similar to (11.13). For a surface that deviates appreciably from the spherical, the effective mass is consequently defined by a very complicated function. Polyvalent metals furthermore have Fermi surfaces in several zones, and an "effective mass tensor" must be defined for each sheet.

For polyvalent metals consequently the concept of effective mass is much less useful than for semiconductors. However, a not dissimilar simplification is often useful. This is the "two-band approximation", in which the carriers are separated into two groups, each characterised by a different scalar effective mass. At first, it seems surprising that so crude a simplification should be useful, but in fact the model provides three independent adjustable parameters—two effective masses and a partition factor dividing the carriers between the two bands. These three parameters are sufficient to permit the model to fit a number of experimental observations, but it should be stressed that the model is purely empirical; it bears only a very slight resemblance to the actual situation in a metal.

The concept of a sharp Fermi surface is, of course, justified for a pure metal at all temperatures, since the valence electrons must constitute a very highly degenerate fermion gas. It is not, however, so clear that either the band picture or the free quasi-particle concept can be justified in general for metals. So far, the band theory has been justified only for extreme situations: it is first necessary to demonstrate that it has a general validity independent of the nature of the binding of the valence electrons and the potential in which they move. In particular, use of free electron expressions, even as modified by the effective mass tensor, would seem to be essentially a restatement of the Sommerfeld model, with its attendant difficulty in accounting for the observed temperature dependent resistivity of a metal. It is also by no means clear why it is possible virtually to ignore the ion cores, although they compose a not insignificant fraction of the total volume of the metal.

The resolution of all these difficulties is due to Bloch, who in 1928 modified a well-known theorem of pure mathematics originally proposed by Floquet. In the form due to Bloch, this theorem in one dimension states that the solutions of the equation

$$\frac{\partial^2 \psi}{\partial x^2} + \frac{2m}{\hbar^2}[E + V(x)]\psi = 0 \qquad (12.37a)$$

where $V(x)$ is periodic in x with period a; i.e.

$$V(x) = V(x+a) \qquad (12.37b)$$

have the form

$$\psi = u_k(x) \exp(\pm ikx) \qquad (12.38a)$$

where the function $u_k(x)$ is periodic in x with the same period as V: i.e.

$$u_k(x) = u_k(x+a). \qquad (12.38b)$$

This result means that the Schrödinger equation for a particle moving in a periodic potential field has eigenfunctions which are plane waves modified by a modulating function with the same period as that of the lattice. It should in particular be noted that irrespective of the form of u the wave functions are not bounded: it follows that the electrons are not in principle localised. Furthermore, for the special case where u is not strongly dependent on the wave number k the solution (12.38) is very similar to the free electron function (5.4). These results indicate why the electrons may be regarded as free in solids, since wave packets describing the electrons may be built up from eigenfunctions having the form of plane waves. If the band is full, of course, as one constructs one packet, one automatically constructs its mirror image, so that no net transport can take place in a full band.

It may also be demonstrated as a corollary of this theory that for some ranges of the energy E the wave number k is wholly real, while for all other values of E, k is wholly imaginary. Further, as E increases, regions of real k and imaginary k alternate. Since it has already been demonstrated that eigenfunctions with imaginary k are not well behaved, and correspond only to particles incident on the region of interest from the outside and in the process of being reflected out again, ranges of the energy corresponding to imaginary values of the wave number are consequently "forbidden". Accordingly, the energy eigenvalue spectrum consists of alternating permitted and forbidden bands. Finally, it can be demonstrated that the total number of states in a band is equal to the total number of cells in the potential V. All these results may, of course, be easily generalised to three dimensions. It can be seen that by producing these results Bloch has demonstrated that the principal features of the band model are *necessary consequences* of the periodic nature of the potential in a solid, and do not depend on any assumptions made about the *form* of the periodic potential, or indeed on any assumptions concerning the nature of the binding of the electron. However, it should be remembered that the potential experienced by an electron in a solid is only periodic in the one-electron model: plainly, if one also takes account of the potential due to the other (mobile) electrons, the total potential will not in general be periodic. The Bloch theorem, like the other models used in this book, is strictly a one-electron model.

Bloch's model shows that the free-electron behaviour, and the lack of resistivity, is a property of a periodic potential. In particular, an electron moving in a perfectly periodic potential will experience no resistance at all. This is compatible with the observation that very pure metals at very low temperatures do in fact have very low resistivities, and very high mean free paths. However, as the potential departs from perfect periodicity, either as a result of the vibrations of the ion cores as the temperature rises, or due to impurities, dislocations, point defects, or stacking faults, the resistivity increases, and the mean free path falls. Since the departure from periodicity due to an impurity or a defect is largely geometric in character it is, to a first approximation, independent of temperature (Matthiesen's rule). The departures from periodicity due to lattice vibrations however plainly increase with temperature, accounting for the positive temperature coefficient of resistivity which is usually regarded as the principal characteristic of a metal. In terms of the quasi-particle notation we have used in this book the scattering of the wave packets by the lattice vibrations which produce resistance may be regarded as "electron–phonon collisions". The collisions result in only small energy changes in the electrons, because of the smallness of the phonon energies in comparison to the Fermi energy. In general, the collision simply moves the electron from one part of the Fermi surface to another. There are three possibilities: first, the two points may lie on the same surface in the same zone; second, the two points may lie on the same surface but in different zones; and third, the two points may lie on different surfaces in the same zone. The first example corresponds to what has been called a "normal process" in the discussion of phonon–phonon collisions, and the second corresponds to an "umklapp process". The third case is called an interband transition, and it is only possible in metals in which there are two partly filled bands. The most notable examples are the transition metals, which have an *s* band with low effective mass, in which most of the charge transport takes place, and a *d* band which is very narrow indicating a high effective mass (electrons more localised around atoms) in which consequently not much charge is transported. In these metals a significant contribution to the resistivity is due to electron–phonon collisions in which the electron is transferred from the *s* band to the *d* band.

A detailed transport theory of metals can be developed on this basis, but it is rather complicated and certainly beyond the scope of the present treatment. The important point to remember is that resistance as such in metals is due to the scattering of the electrons at the Fermi surface by departures from periodicity in the lattice. In semiconductors, on the other hand, the conductivity is more closely connected with the number of carriers, and depends rather less on the collisions experienced by each carrier.

Index

Page reference in **bold figures** denotes reference to table on that page

Printed in Dunstable, United Kingdom

91338432R00147